PREFAB

adaptable, modular, dismountable, light, mobile architecture

PREFAB

adaptable, modular, dismountable, light, mobile architecture

Publisher: **Paco Asensio**

Editorial coordination and text: **Alejandro Bahamón**

English translation: **Bill Bain**

Art director: **Mireia Casanovas Soley**

Documentation: **Marta Casado Lorenzo**

First published in 2002 by LOFT and HBI,
an imprint of HarperCollins Publishers
10 East 53rd St. New York, NY 10022-5299

HBI ISBN: 0-06-051358-6

Distributed in the U.S. and Canada by
Watson-Guptill Publications
770 Broadway New York, NY 10003-9595
Telephone: (800) 451-1741 or (732) 363-4511 in NJ, AK, HI Fax: (732) 363-0338

D.L.: B-34.158-02

Distributed throughout the rest of the world by
HarperCollins International
10 East 53rd St. New York, NY 10022-5299
Fax: (212) 207-7654

Printed by:
Gráficas Anman, Sabadell, Spain
August 2002

If you would like to suggest projects for inclusion in our next volumes, please e-mail details to us at: loft@loftpublications.com.

The term prefabricated brings to mind a building system in which the essential pieces of a structure are sent to the site on which the finished edifice will be constructed partially or completely assembled. Once there, it is necessary only to join and anchor the parts. Although this is still the basic principle of prefabrications, owing to the evolution and the very characteristics of this system, prefabrication in architecture today gives speaks rather of a series of highly varied conditions that bring us new possibilities of how to inhabit our world. Prefab construction is no longer understood only as a method employed simply as a way of saving time and money on a project. It is also an alternative that can solve the most complicated structural situations, including fold-up and fold-down offices, family growth, refuges for inveterate travelers and itinerant exhibits.

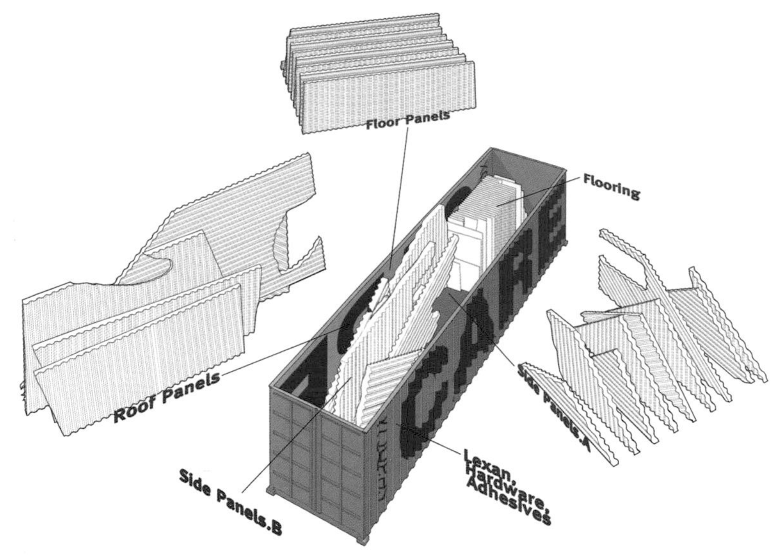

The characteristics found in prefabricated structures have recurrent themes developed to solve such diverse questions as the cost-effectiveness of the operation, flexibility of use, or the mobility of the piece. From the time of the most primitive constructions, such as the Bedouin tents still used today, to the most complex contemporary solutions, e.g., hydraulic structures in stadiums and coliseums, architecture has sought solutions to these needs. It is not, however, until recent times that technological advances have allowed us to manipulate very lightweight materials and simplified construction techniques, thus enabling these models to find a sophisticated execution that adapts to the complex needs of contemporary life.

Although mobile edifices are found among early human artifacts, the development of this mode of construction is strictly linked to technological advances and the discovery of new materials. In this sense, then, universal expositions and world's fairs showcases for the latest technological trends have always served as the ideal events for the creation of new lightweight and easily dismantled structures that are still capable of standing up to the intense use of thousands of visitors. It is in this field that we have been able to test the capacity of the building arts to create an adaptable object that takes concepts from engineering, industrial design, and other arts. Out of these universal expositions have come a wide range of transient structures. Examples abound that range from universally acknowledged urban icons to domestic residences that have remained with us to the present day. The Eiffel Tower, perhaps the definitive emblem of this type of constructions, has been standing since it was raised for the Paris Exposition of 1889. And it has served as an example for decades of the building properties inherent in metal. The Norway Pavilion for the 1929 Exposition in Barcelona, a relatively unknown example, was bought and transported to the mountains and is transpose used today as a private second residence.

In the twentieth century, after the popular development of construction with prefabricated concrete panels and wooden modular houses, this type of building system came to be considered as a solution to problems brought about by population explosions in larger cities. At the same time, however, it was very limited in design terms. Prefabrication, because of its formal and functional limitations, tended to leave out particular needs for same client's, or imposed specific conditions involving the placement of the building. Rigid prefabricated pieces, the basis of these constructions, prevent the development of designs. It was difficult, under these conditions, to be able to many the construction to environmental concerns of sunlight, ventilation, topography, or given urban elements. Thanks to recent advances in the fields of design, construction methodology, and materials, this type of architecture is now able to handle sensitive solutions for urban and natural environments. As a result, a serial architecture is now possible, both in terms of the systems that control design and the techniques used by construction. It responds to the particular needs of each client and to the particular conditions of the site on which it is built.

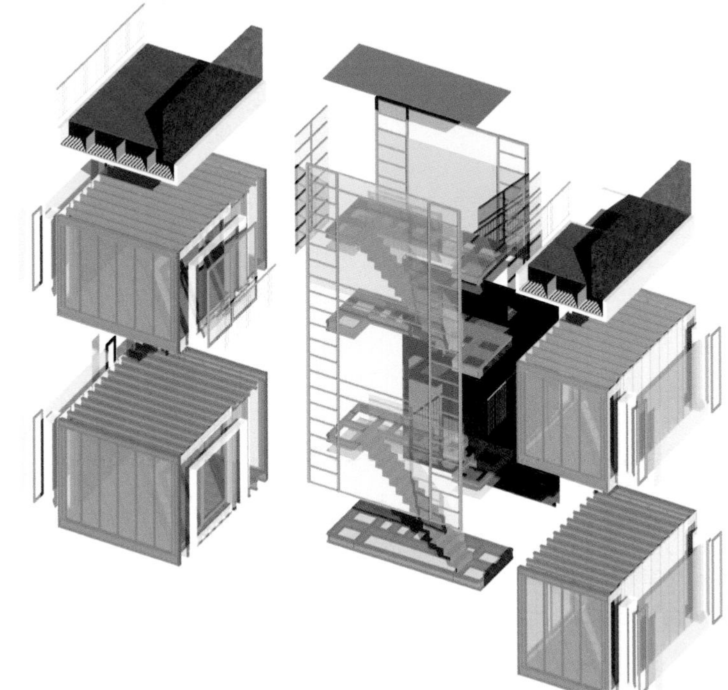

Nowadays, prefabrication no longer reguires situations as unique as passing event or a solution for a whole neighborhood of row houses. Architecture with a prefabricated core is being used to create a wide range of objects, many of which are very ambitious in terms of size or flaunt their own special construction. Projects involving modular buildings that can cover enormous tracts of city, such as the Wooden modules project developed by Arthur Collin in East London, or floating structures that create an artificial landscape in the middle of the ocean, like the Floating Island by Softroom Architects are constructions that are so flexible that, at least in its current state, static architecture would find impossible to equal.

The specific concepts on which prefabrication prides itself are those that define this compendium. These works are designated by characteristic features like the lightness of the buildings of Jean François Revert for the Keroman Base, comprised of polycarbonate panels, or the mobility of the structures designed by Joost Glissenaar and Klaas van der Moelen, using a transport container as a take-off point to develop a lending-library for children's toys. These features are used to organize this book into five building typologies: mobile, adaptable, lightweight, modular, and dismantling. All of the examples in these pages share some or all of these features, but each project is, at its core, most striking because of one particular trait. Through this survey of constructed pieces and projects in blueprint, the range and the skills shared by today's designers, builders, and manufacturers of prefabricated building materials is showcased.

Dismountable
Mobile
Light

Adaptable
Modular
ZIMMEREI
FRANK

Duimdrop

Joost Glissenaar + Klaas van der Molen / BAR

This object, which may at first sight seem enigmatic and hermetic, sets up a relationship with the environment into which it is introduced. As described by its architects, Duimdrop "is there to create a new McDonald's of social renovation in Holland". Into the context of public squares and parks where these containers are introduced, they might appear to be autonomous object but their use is totally bound up in the immediate environment.

The box, the most rudimentary of architectural forms, is a kind of warehouse and distribution center for the playthings of local children between four and eleven years old, the majority of whom are immigrants. Since the parents of these kids usually send money to their home countries of origin, toys don't get top priority in their family environment. Duimdrop is slated to alleviate this situation and to create at least one small play space for the smallest children.

The basic toys are dealt with and lent out by someone who takes care of administration for the box at given scheduled hours. The most popular of the playthings are lent out for "Duimdrop money", which can be earned by doing little chores around the neighborhood. The system thus not only helps solve the community's recreational and educational needs, it also creates a dynamic to improve relations among neighbors. Duimdrop serves to better communities using a prefabricated approach.

Architect: **Joost Glissenaar + Klaas van der Molen / BAR**
Location: **Various**
Surface area: **161 sq. feet**
Construction date: **2002 (last model produced)**
Photography: **Rob't Hart**

Adaptable Modular Dismountable Light Mobile

The developers of the project, a group of social workers, were initially in favor of an organic form for the structure, perhaps along the lines of an amusement park reference. Later, given the fact that Duimdrop would be distributed in settings where urban and social make-ups vary substantially, the preference changed into an abstract shape that would make a neutral statement in regard to its placement.

The containers used are those which, instead of sheet metal, use a wood veneer paneling for the exterior decoration. The material is an easily perforated surface that makes it easy to install windows and doors. Originally, the boxes had apertures like those used in storage equipment. But with the product's success, containers were made to order for these cabins based on the type of framing available on the market.

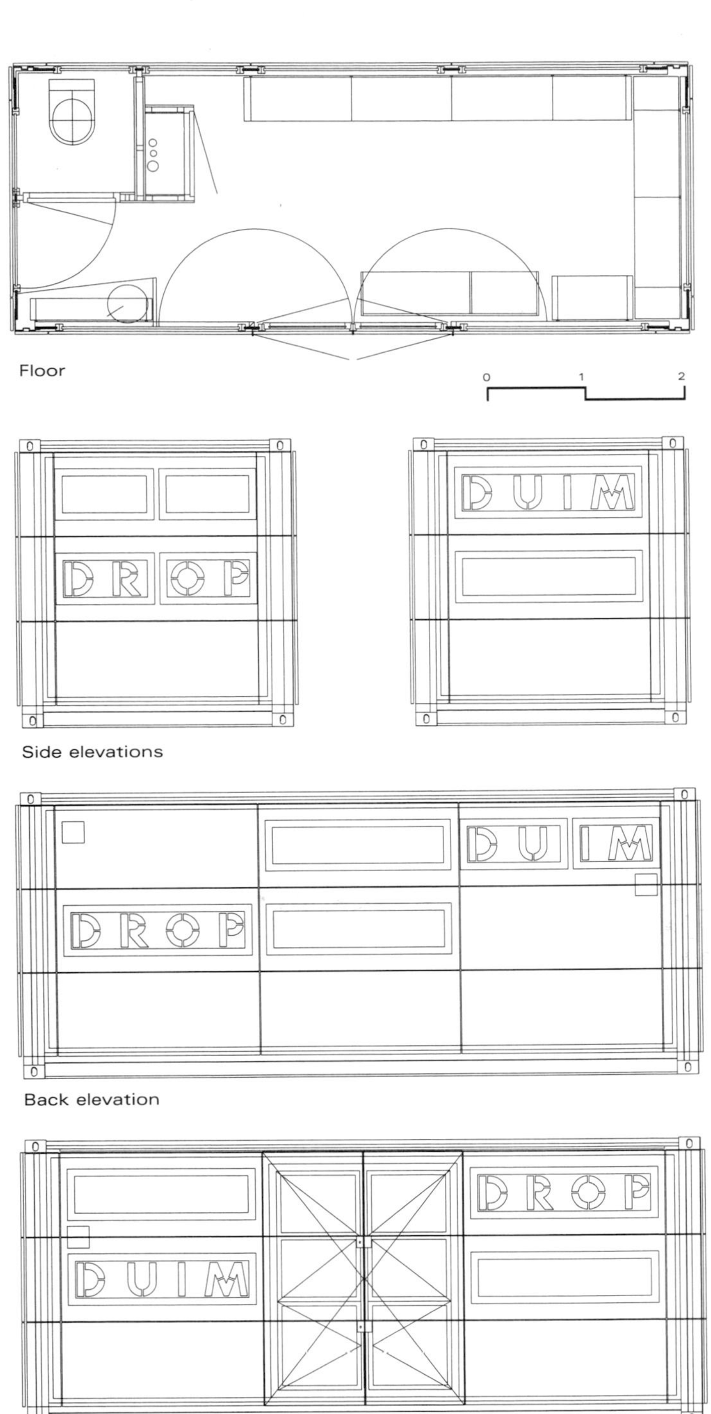

Floor

Side elevations

Back elevation

Front elevation

Section

The small proportions of the container, acting in combination with the colors and the enigmatic look, turn the object into a piece of urban furniture or, with windows and doors closed, a sculpture.

Walls and doors: These elements, and the roof, are thermally insulated from inside to deal with temperature extremes. The finishings are a wood veneer painted yellow. The container consists of a small administrative office, a lavatory, and a storage area for the toys.

A second coat of metallic mesh, primarily used in emergency stairways, is applied onto the container's outer panels. This mesh serves the dual purpose of protecting the cabin from vandalism (mostly graffiti) and also of creating greater scale, detail, and depth while maintaining its durability and resistance properties.

The windows, the doors, and the signs are holes in the veneer. Permanent ventilation is thus served as well as lighting. The artificial light that fills out the natural illumination accentuates these openings and acts as city street lighting to add to ordinary security. The box thus is made over from a lifeless object to a mysterious magic trunk with playthings in it.

Kielder Belvedere

Softroom Architects

Softroom Architects' were charged with building a small refuge in Kielder, one of the largest forests in the United Kingdom. The result is a belvedere that comprises the first architectural commission in an extensive public art program that was developed by the Kielder Society over the course of the last three years. Following a project to promote sculptures, a limited competition was announced, calling for designs for a refuge on the northern edge of the reserve.

The design responds to the Society's needs and to the severe climatic conditions of the region. After ten months of design and construction, the structure is used now as a shelter for park visitors. During the spring to fall months, the belvedere accommodates passengers waiting to board the ferry to cross the lake.

The local weather conditions called for the use of pretty heavy duty materials. High resistance was needed in order to last the destres lifetime of at least ten years. The chairs are made of resin and are thus practically indestructible. The wall screens are covered in stainless sheet steel, giving the structure a less chunky look.

Architect: **Softroom Architects**
Location: **Kielder, U.K.**
Surface area: **129 sq. feet**
Date: **1999**
Photography: **Keith Paisley, Forum (for photographs of the construction)**

Kielder Belvedere exhibits the wide panoramic views of the human-made landscape. The steel sides of the refuge reflect a partial image of the forest surrounding it. The view in the distance across the lake is framed and reflected inside the structure by the polished convex mirror surfaces in the front.

A sensation of warmth and comfort has been achieved by finishing the inner walls with gold-colored dust and installing lighting in the same tones. The shelter is entered by way of a metallic ramp that connects the slope of the terrain with the floor of the refuge. The slight difference in grade is negotiated by four metal columns that minimize the step up.

Here the built object becomes an exercise that blends sculpture and architecture. The entire proposal is a suggestive and voluptuous way of creating an interesting dialogue with the environment. The design and construction are generated out of modules that plasticize the frame and the skin, incorporating the latest building styles.

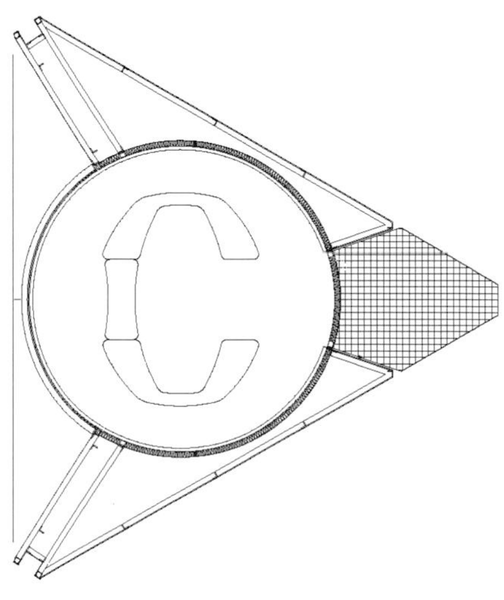

Interior floor

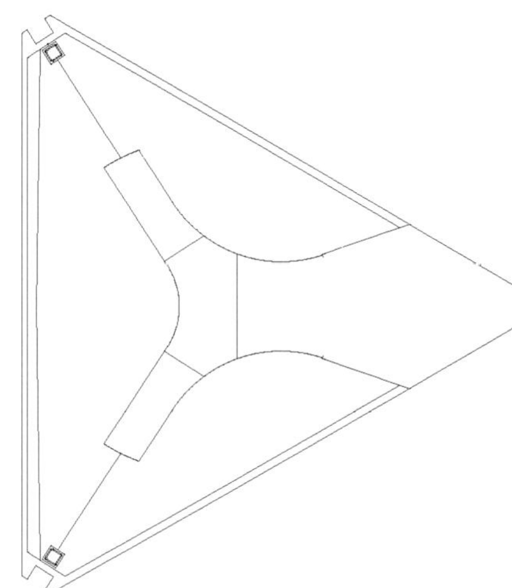

Roof plan

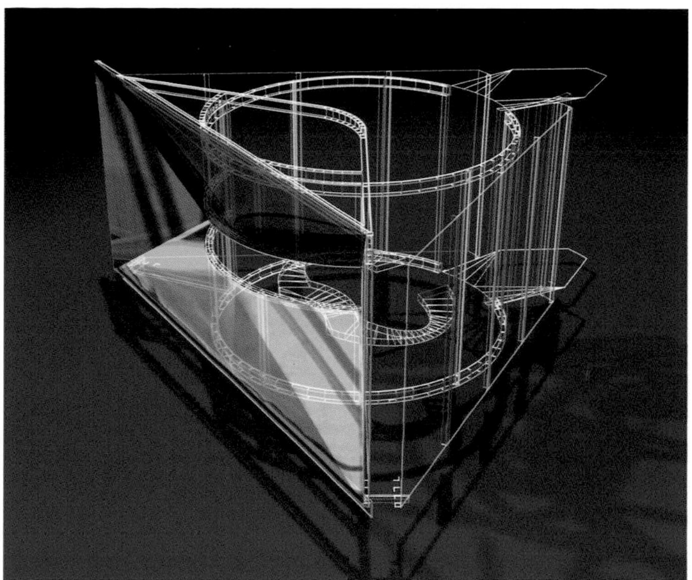

3D model

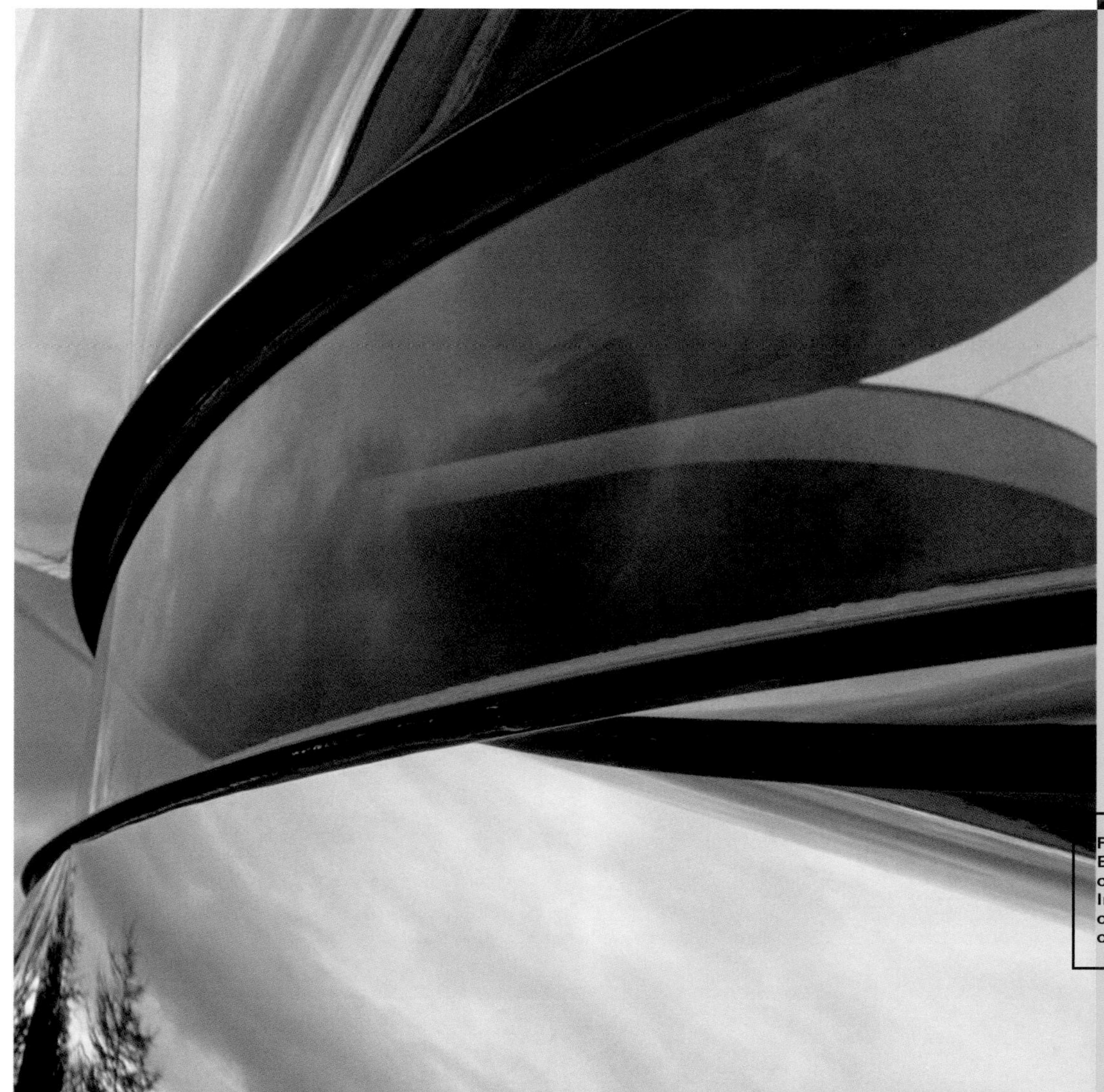

From the outside, Kielder Belvedere appears to be a closed, enigmatic triangle. Inside, we see a circular chamber, warm and welcoming to weary walkers.

Modular Dwellings

Edgar Blazona

The origin of this project is found in the angst of furniture designer Edgar Blazona. Interested in modern architecture, he has explored the aesthetic of graffiti and of skaters. He considers himself an urban artist, and his name is not unknown in San Francisco for his work on clubs and large music festivals. Now, after working for a number of design firms and creating his own design studio, he has conceived, designed, and constructed this project, called Vivienda Dwellings.

Modular Dwellings flexibility, in comparison with other traditional prefab home systems, lies in its ease of assembly. This characteristic allows the consumer to put it up, take it down, then transport it easily at reduced cost. This flexibility is possible thanks to the use of materials like galvanized sheet metal, corrugated metal, and fiber glass panels. Sixteen cement bases serve as general anchoring and leveling elements. They are erected onto the eastern side of the site some 7.8 inches (20 cm) above grade. This avoids water filtration and causes less damage to the topsoil. With this base, a simple cube is put up using plywood panels for the floor and sheet metal for the walls and roof. The open and transparent side of the refuge is fiberglass.

Designer: **Edgar Blazona**
Location: **Mobile**
Surface area: **161 sq. feet (extendable)**
Construction date: **2001**
Photography: **Glen Campbell, Julia Blazona**

The designer's goat is after with this project-which combines interests in art with bold and daring design concepts-is the creation of so-called guerilla architecture. Blazona defines it as mobile, easy edifices that assemble and serve anytime and anywhere. It is a design logic that employs the most basic materials instead of complicated and sophisticated costly facings. Modular Dwellings is for those who need to put more space in their life but still value design and comfort over commercial conveniences. Mobile buildings. That is the simplicity of it: they are modern, they are cost=effective, and they adapt to any environment.

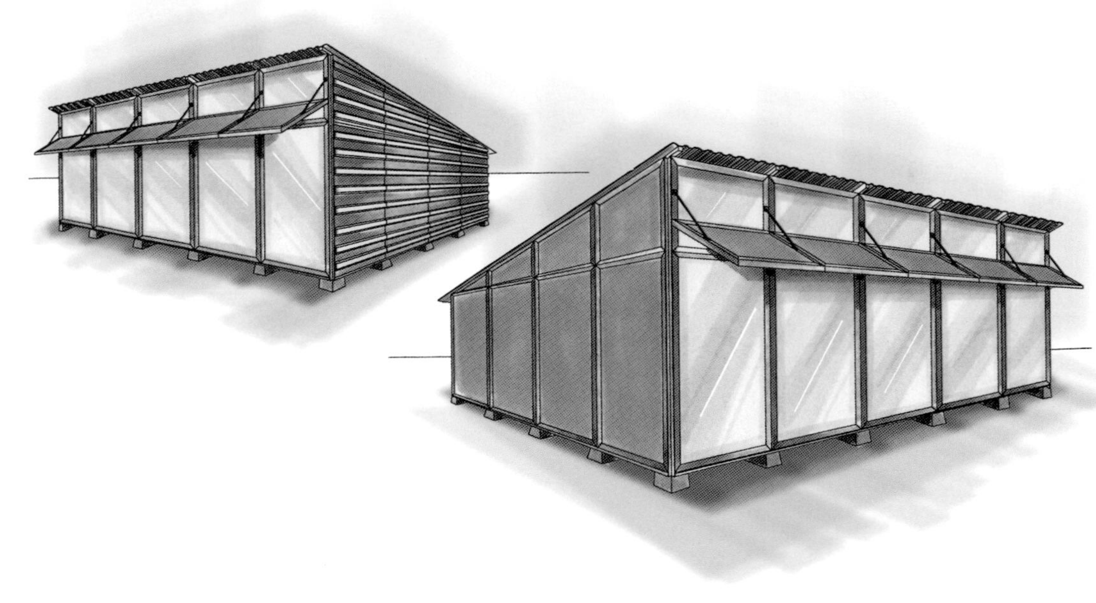

Black Maria Hiroshi Nakao

This unique object is in a changing modular system halfway between a small pavilion and a large-format sculpture. It was designed by architect Hiroshi Nakao for the exhibition "Art Today" at the Sezon Museum of Modern Art, in Japan. Installed inside dense native woodland of the region, Black Maria creates a dialogue between landscape and built object by exploring the different possibilities in which it can be arranged.

With its curved walls, the object suggests that it could be a variety of different thicknesses. When it is completely closed, the composition creates a deep black empty space. Slowly opened, the vacuum disappears and a large hole is produced that frames different views of the woods. When the opening reaches its maximum extension, the hole disappears again and the original emptiness doubles over on itself, like a glove turned inside out.

It is a folding screen, but unlike the conventional screens that divide a space, this one breathes space in and out of itself. In the midst of this suggestive game of space, form and surroundings, the alternatives of different distributions let you use the object in different combinations as the landscape's objet d'art, small pavilion, covered bench, "room" divider etc.

Architect: **Hiroshi Nakao**
Associate architect: **Hiroko Serizawa**
Location: **Sezon Museum of Modern Art, Karuizawa, Japan**
Surface area: **215 sq. feet (minimum)**
Date: **1994**
Photography: **Nacása & Partners**

Black Maria comes from the nickname given the film studio designed by W.K.L. Dickson and build in 1893 for Kinetoscope, a room invented by Thomas Edison to watch a moving picture through a sight device. It was a black box, like a one-family home of the time, or a black American police van. The roof folded to let in natural light, and as the structure was on piles, it was possible to move to find the best position with respect to the light. Hiroshi Nakao's project takes these references and makes them over into a reflection on the desire to control light and detain movement.

The Black Maria is made of a wood frame with pine plywood panels. All of the wooden parts going into the assembly of the little building are treated with black oil to achieve the desired color. At the same time, this treatment seals the material's pores and makes it much more resistant to the climate. Each module is connected to the next by way of large metal hinges that allow its transformation.

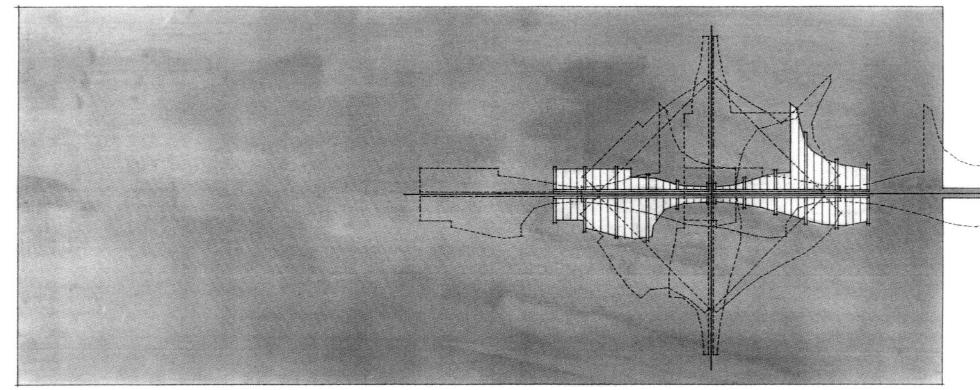

Movement schemes

The sinuosity of the cladding is carefully designed to be able to fit into its own elements in a variety of different positions. The fine metal pillar arrangement in the walls contrasts with the capricious shapes that make up the perimeter.

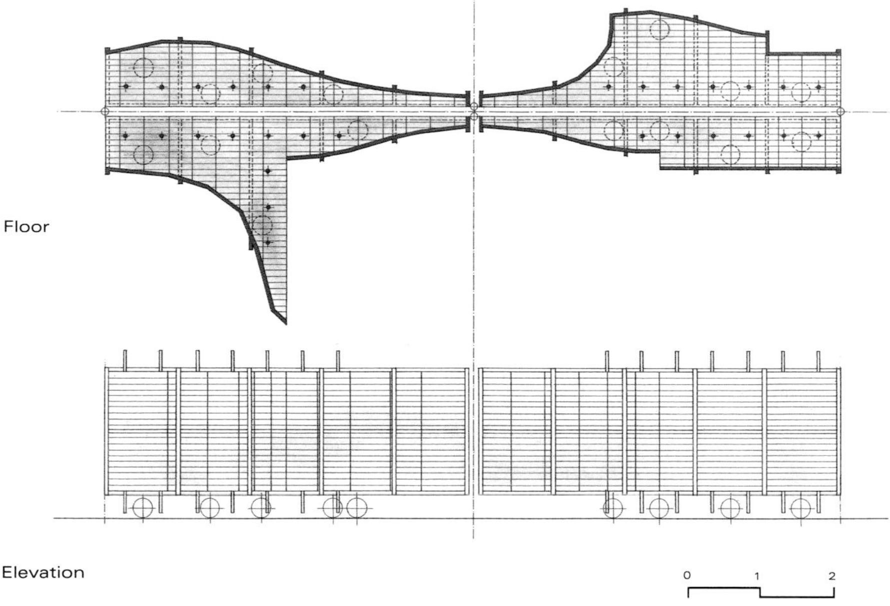

Floor

Elevation

This wooden structure is reinforced by metal pieces that secure the object from above and from below. The wheels beneath it are anchored to the two layers of plywood; they make it possible to negotiate the uneven terrain. Inside, a series of fine metallic elements help support the roof, creating an artificial interior woodland that is dispersed in the surrounding landscape.

Wheel detail

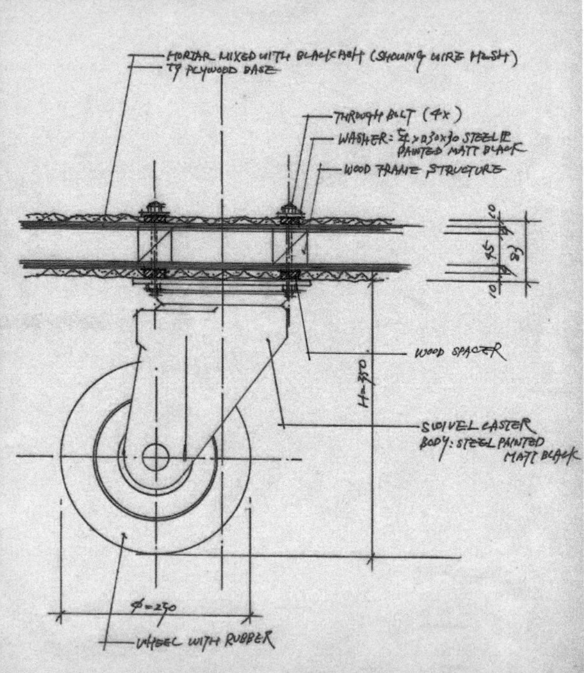

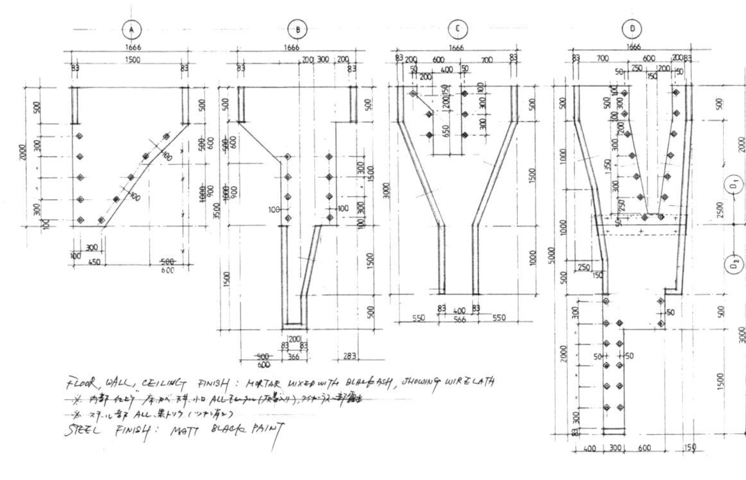

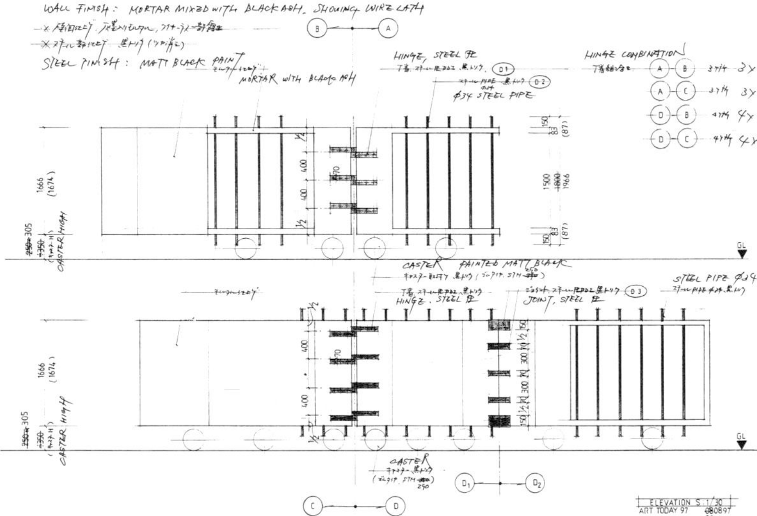

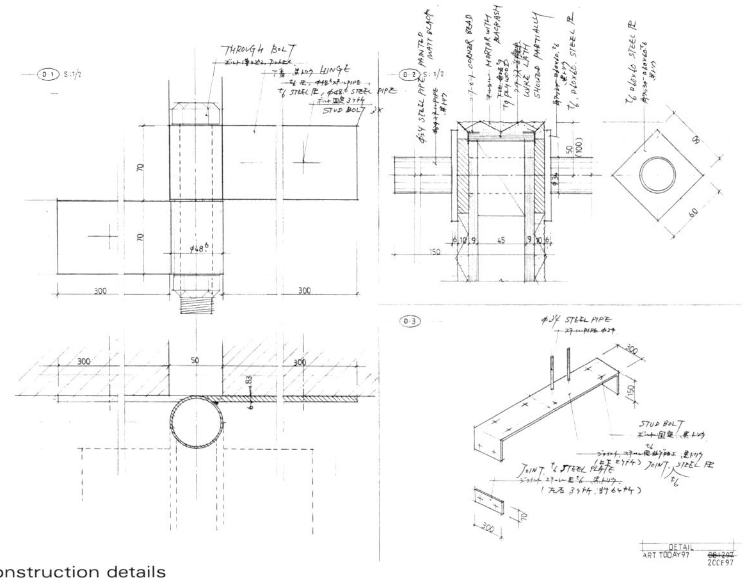

Construction details

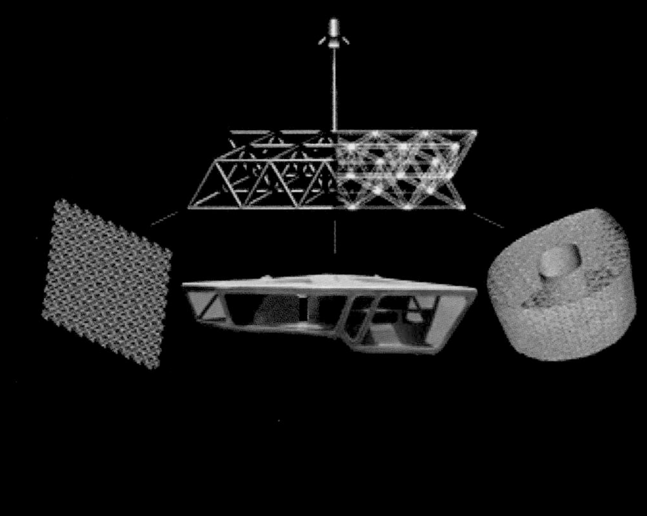

The spirit of the times, when movement and occupation of the farthest corners of the Earth seem to be the word, launched the development of this project.

Architect: **Meindert Versteeg** Location: **Mobile**
Surface area: **323 sq. feet** Construction date: **1999**

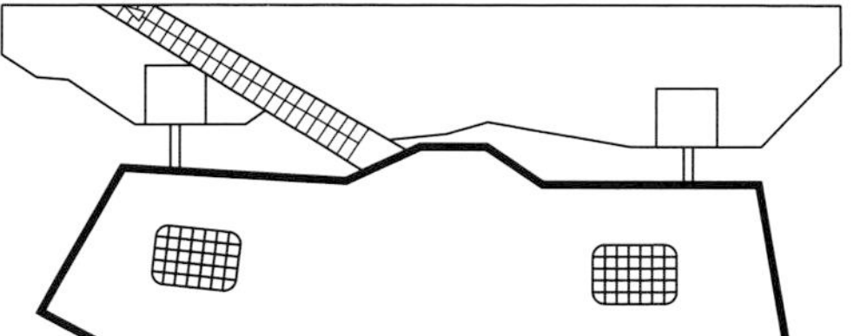

Cliff House

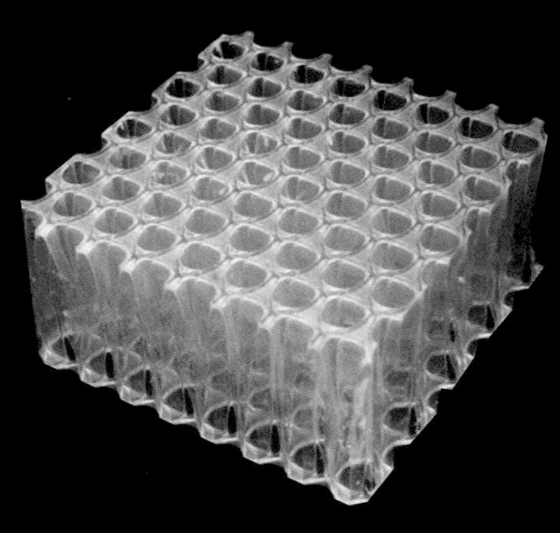

The main interests of young dutch architect Meindert Versteeg are windsurfing and snowboarding as well as a fascination with combining materials, lightness, mobility, and naturalness. All these concepts he constantly attempts to integrate into his projects.

The Cliff House project addresses all of these concerns. It consists of a transformable space, not only able to be slotted into any spot for any length of time, but also because of its functional and formal features, adaptable to needs that may arise. The unit is a transparent shell that has the distinct advantage of being re-locatable and highly flexible. The ease of transformation depends on the number of people living in Cliff House, the season of the year, or the purpose of its operation: house, café, bar, research center, etc.

The study of new materials applicable in the unit makes it possible to realize a comfortable space with a surprising panorama. The interior is provided with beds, chairs, and work tables, in aluminum derivatives. The floor is constructed from the storage crates used to shipped the materials that build the cabin. Depending on the use chosen for it, the unit can even incorporate a small bath. Water is collected in tanks in the structure's frame.

Imobile is mobile work space, light and flexible. It it is an asset into a society in a permanent state of change.

Architect: **Jennifer Siegal / Office of Mobile Design** Collaborators: **Elmer Barco, Arona Witte, Ashley Moore, Saul Diaz, Jason Panneton** Location: **Mobile** Surface area: **150 sq. feet** Construction date: **2000** Photographer: **Benny Chan (models and projects)**

iMobile

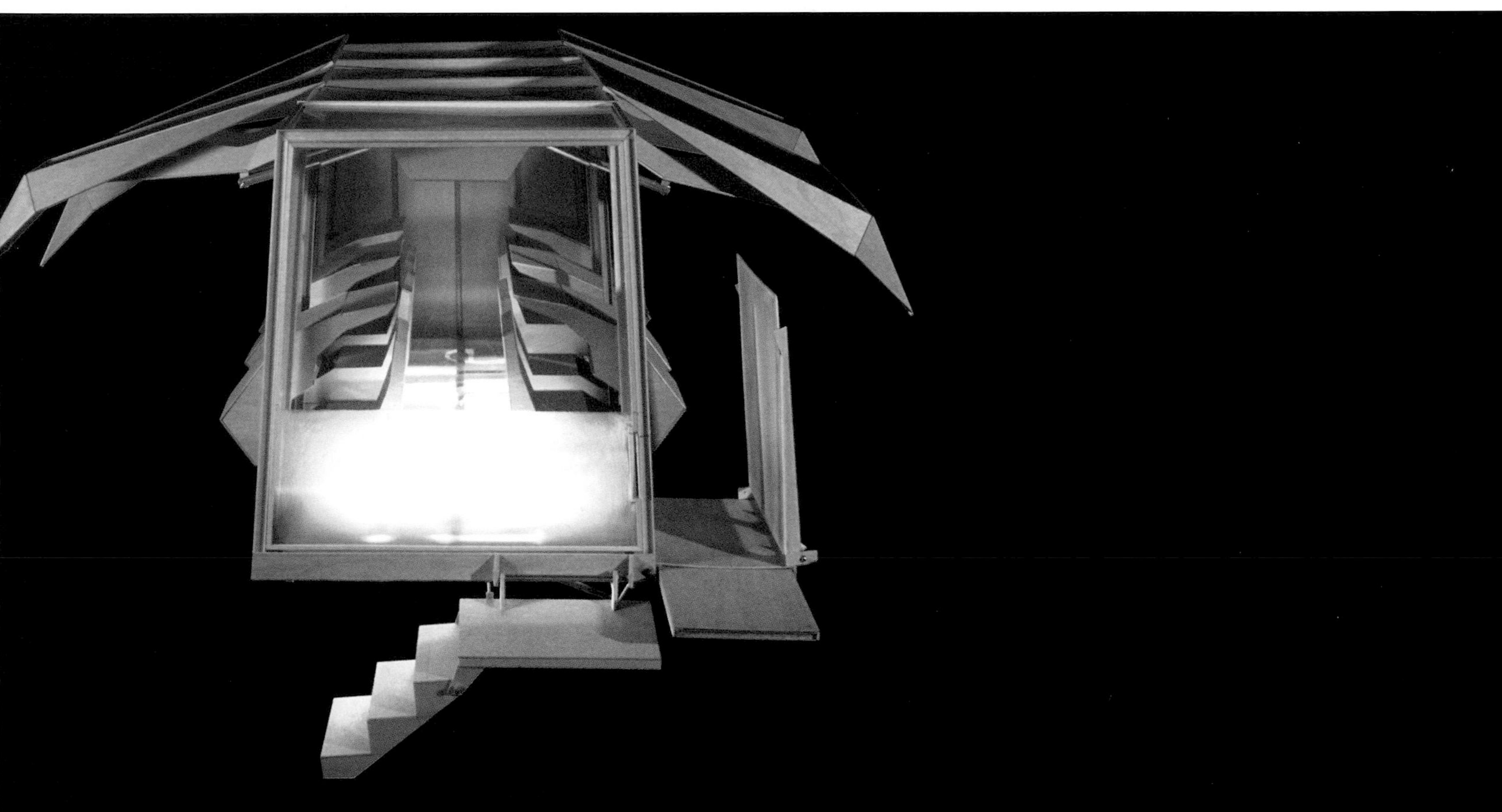

iMobile is a movable permanent connection to global communications networks with an ability to work as the advertising pavilion for the latest systems in information, hardware and software. This is an advance in the concept of mobile and dynamic enterprise; a work space that evolves into the capability to handle business growth, contraction, and metamorphosis.

iMobile uses constructive technical and data processing solutions, for the modern business. Firms requiring an economy of movement will welcome Imobile. This adaptable and flexible framework always responds to its immediate and changing setting.

The piece has to permit flexibility, resistance and durability. What allow these conditions is the use of high quality materials that are also lightweight and cost=effective. The self-sufficiency and ease of transport of the piece takes it anywhere and lets it generate the model for a future city.

An entire artificial ecosystem: an island and a small summer house incorporated into a carbon fiber bubble that can be rooted anywhere.

Architect: **Softroom Architects** Location: **Mobile**
Surface area: **700 sq. feet** Construction date: **1997**

Floating Island

Density, the desire for mobility, and the lack of space to build on are themes that affect more than just the permanent home. This group of young architects has created a temporary, portable, and floating proposal: a small refuge that can be transported and anchored on any site. This lightweight structure of carbon fiber is designed to be open, a stabilizing element, with a floor and a terrace for the cabin.

The furniture inside -the bed, the sofas, and other pieces- inflate as soon as the top of the module is opened to activate a small folding island powered by a generator pump. The project is an artificial landscape that creates the appropriate conditions for this small marine refuge.

Construction of a model of this sort could be expensive, but mass production might mean cost on par with a private yacht. This proposal has allowed the architects to contemplate possible solutions for densely populated waterside cities like London.

Yardbird Studio

Neal R. Deputy Architect Inc.

The Yardbird project is a small office of 269 square feet that can be installed in any backyard. The concept is the creation of a quiet work space away from the other activities of the household while maintaming the advantages of working from home. Combining these needs along with some historical and architectural reference points of the region was the concept behind this basic geometrical object and an elemental structural system.

The Yardbird prototype was constructed in Charlottesville, Virginia, USA, where the client and the architect shared common go a is with regerd to the region's landscape, which to a part of the neoclassical architectural legacy of Thomas Jefferson and the developmental center of modern architecture during the first decades of the twentieth century. Jefferson's reinvention of classicism through Virginia's rural building traditions was the starting point for the design of the Yardbird Studio .

Drawing on classical European models but staying true to local materials like common red brick, wood, and metal roofs, the architects desired to melt the two styles.

Architect: **Neal R. Deputy + Gaither Pratt**
Location: **Charlottesville, Virginia, USA**
Surface area: **269 sq. feet**
Date: **2000**
Photography: **Jim Rounsevell**

To avoid falling into an architectural redundancy by designing another Jefferson's style building just like those around, the project's classical roots were expressed through the use of local materials and building techniques. While this method brought about the purest analytical model in the Jefferson heritage, Yard bird Studio takes aduantage contemporary elements. Witness, for example, the steel window frames, the rough-textured metal facings, or the revealed metallic frame, all of which honor the semi-industrial heritage of the Charlottesville community.

From outside, the project is a sheet metal box raised on svelte metal columns, anchored, in their turn, to a concrete foundation below site grade. The front and side windows use a prefabricated aluminum framing system that opens outward.

Inside, all of the horizontals-the floors, the work tables, even the shelves-are in wood and rubberized veneer, which provide more durability and easy maintenance. The steel stairway, the awnings over the windows, the lighting system, and the storage system come right off the industrial market, allowing the project to be not only economical, but fast and easy to install.

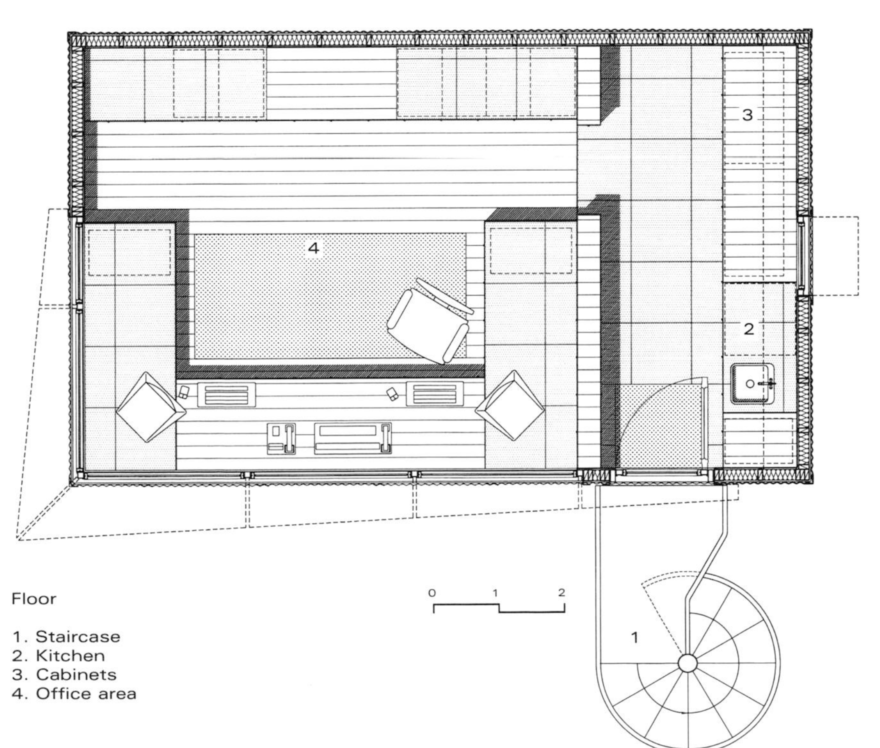

Floor

1. Staircase
2. Kitchen
3. Cabinets
4. Office area

Location plan

The Studio couldn't be simpler in its desingn. It is a room some 64.5 by 37.5 square feet elevated by raw steel columns a small building is based on standard dimensions and prefabricated building materials. This simplicity makes it easy to modify the space as require, from a sweeping, open space to an interlinking set of work modules.

The raised bay makes it possible to use the lower story as desired, perhaps as a gardened area or for parking. At the same time, a independent in work space is gained as well as an attractive panorama. The metal window frame, it should be added, has a visor to protect from direct sunlight.

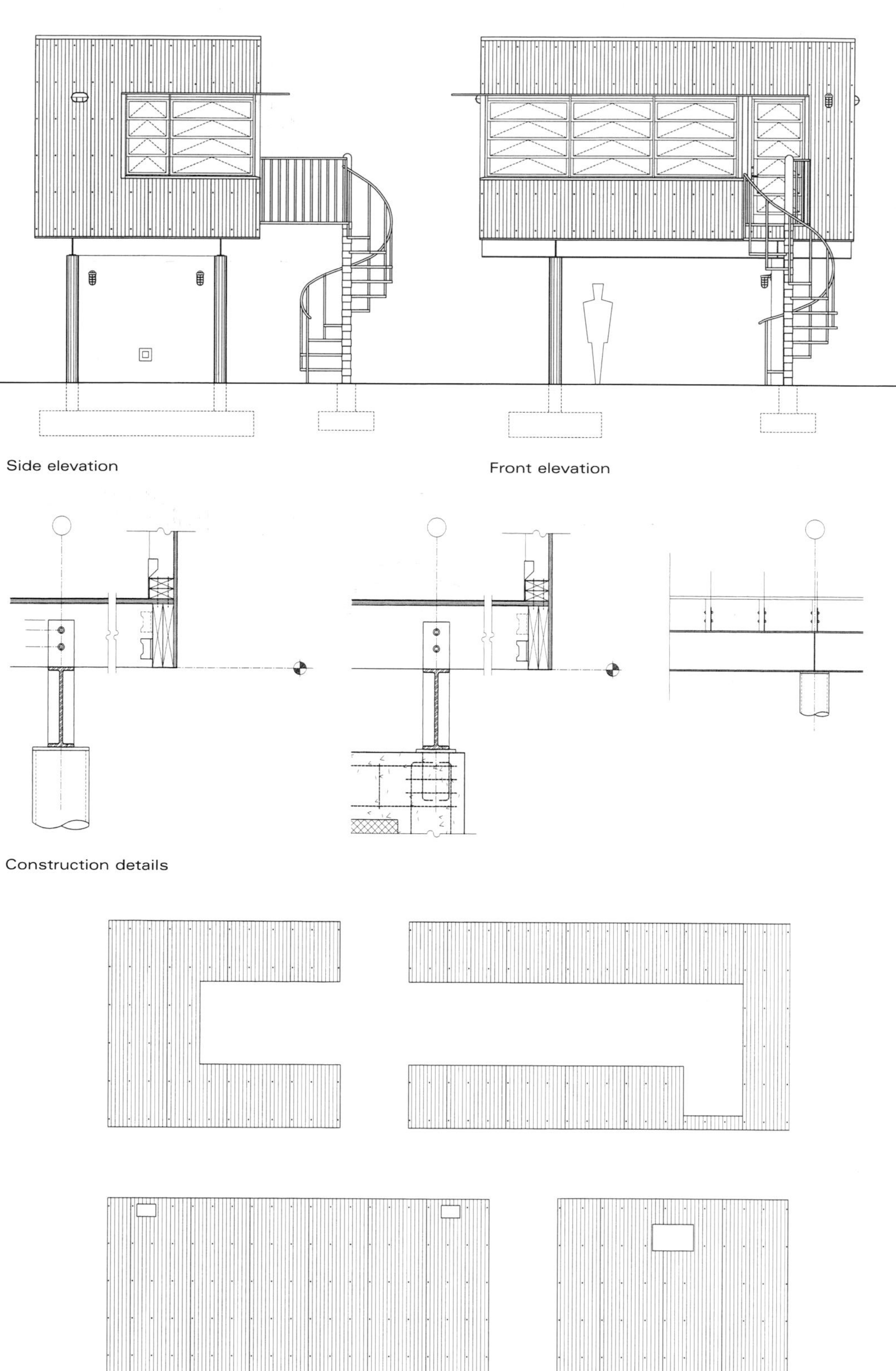

Side elevation

Front elevation

Construction details

Section scheme of metallic sheet

The shape and the dimensions of the space allow excellent interior flexibility. The project's can be adapted to many different needs.

Keroman Nautical Base

Jean-François Revert

The former Keroman submarine base in the northwest of France fell into disuse after World War II, so the city of Lorient, located on the banks of the Ter River, decided to reclaim this area and integrate it into the urban grid and the region's landscape.

The project was conceived around two main premises: to provide a connection to the dominant existing structures, and to provide an incentive to use the area as a public space. The base's original concrete edifices were preserved to retain the traces left by the barbarism of war. Against these enormous structures, a gentle, warm installation of new buildings, smooth, shining, and opalescent, provides a modern contrast to the region's past.

The shipyard is now used for the construction of racing boats, and is also occupied by different nautical service businesses and a public space. The architect proposed two sets of buildings constructed on the riverbanks, next to a walkway that allows the best possible view of these large buildings.

Architect: **Jean-Francois Revert**
Location: **Lorient, France**
Surface area: **35,521 sq. feet**
Construction date: **2001**
Photography: **Stella Rotger**

The project includes a hangar for two America's Cup boats and other smaller sailboats, specialized high technology workshops, services for the boat crews, offices, a research center, and information and communication areas. The boat maintenance facilities, which called for waterborne operation during training periods, played a big part in the organization of the buildings.

The material used for the construction of these buildings is a polycarbonate called Dampalon, which is normally used to improve natural lighting in large structures such as factories, workshops, or gymnasiums. With this material, it is possible to erect a building that is rather lightweight and easy to put together through the use of a metal framework. An ideal interior space is achieved through Dampalon's natural lighting and its properties as a thermal insulator.

Location plan

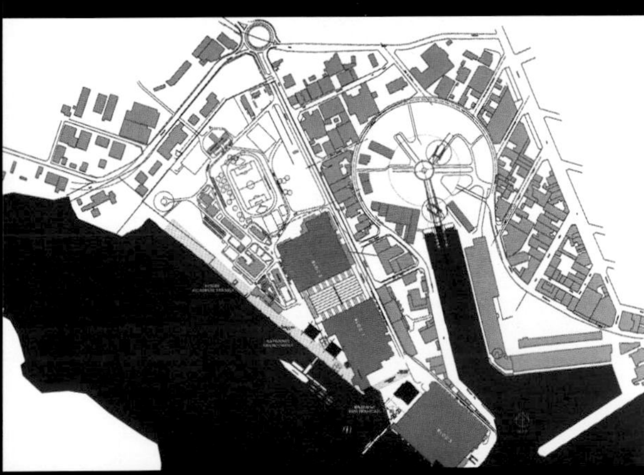

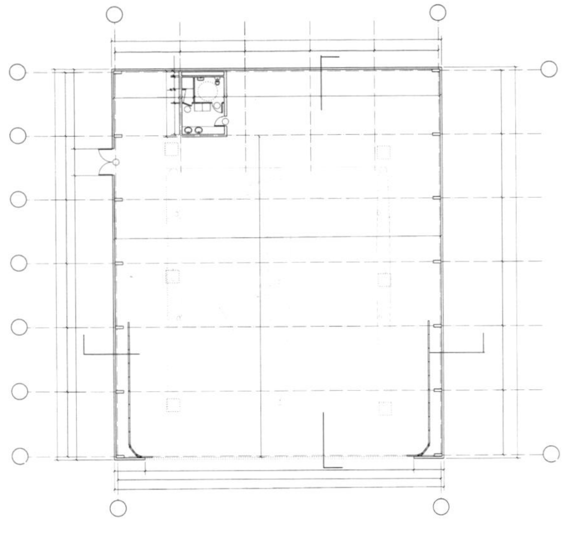

Floor plan for hangars 1 and 2

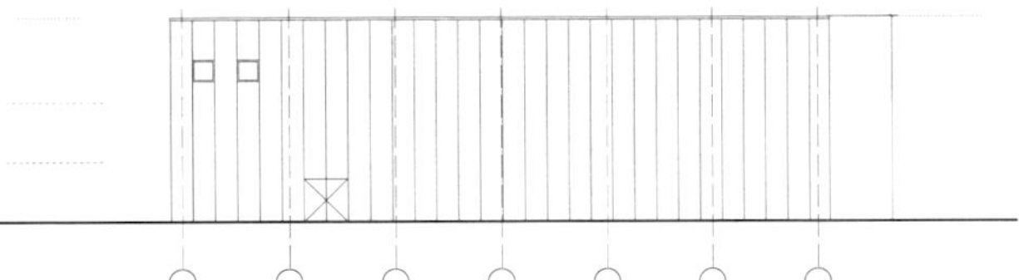

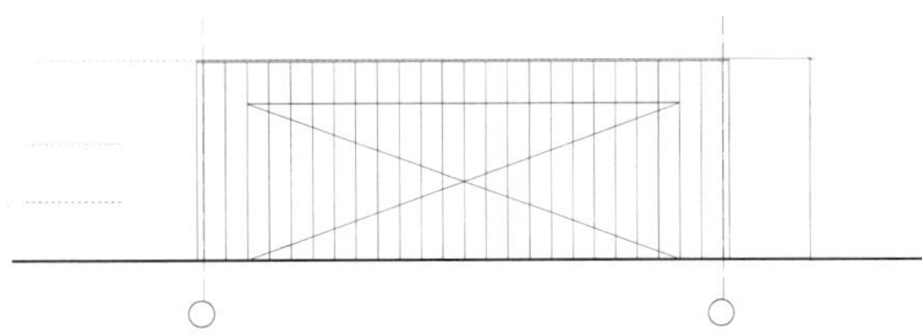

Front elevations

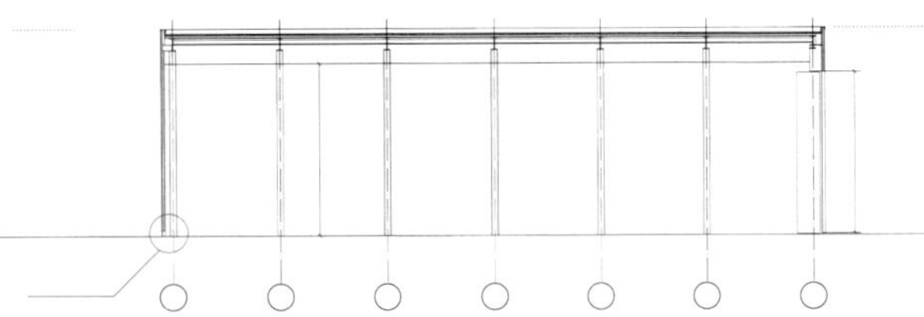

Longitudinal section

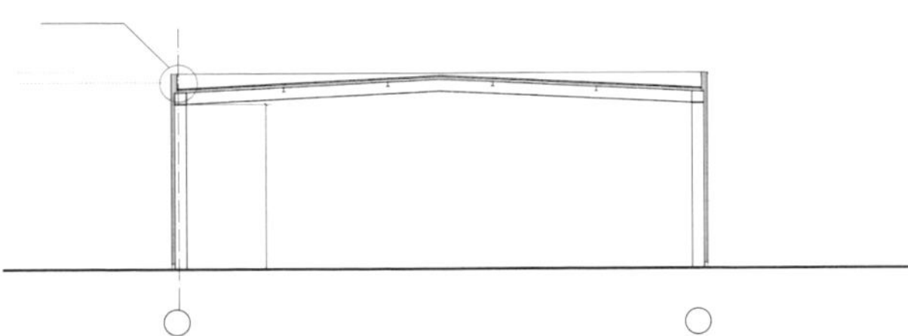

Transversal section

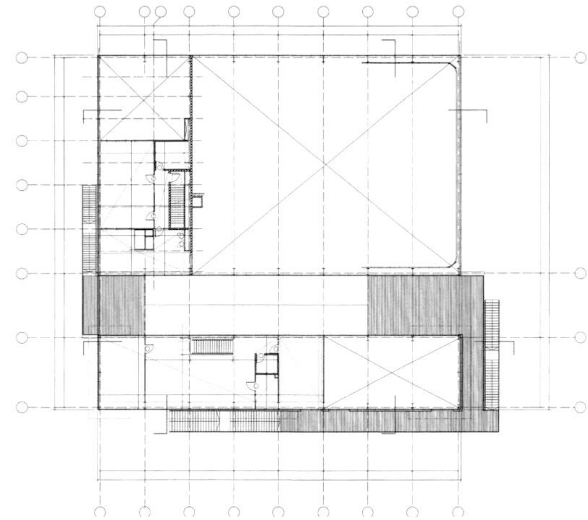

Floor plan for hangar 3

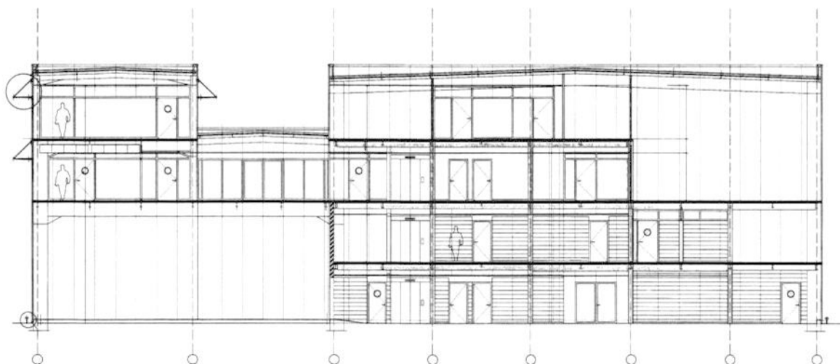

Transversal section

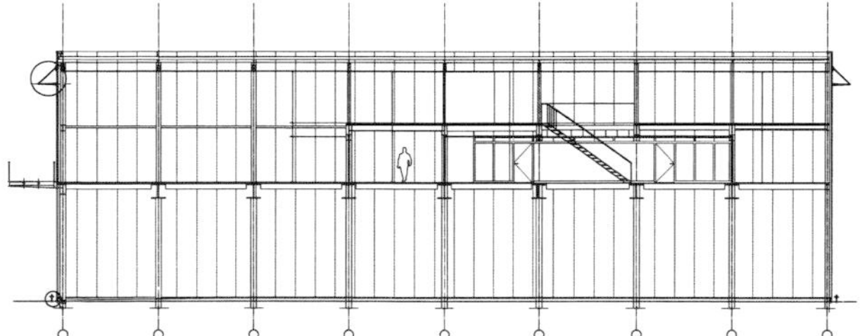

Longitudinal section

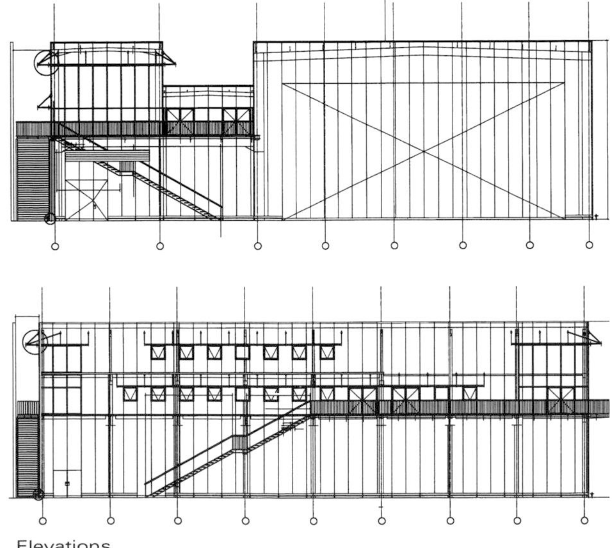

Elevations

The lightweight construction is also obvious in the interior, where the metallic framework, the polycarbonate shell, and the zinc roof are all fully visible. The large translucent surfaces can be moved on rails that accommodate sliding doors so that the hangar is directly connected to the launching platform. Functional windows in the polycarbonate provide natural ventilation when the doors have to remain closed.

The buildings, and the ships built inside them, make up much of the renovation of this public space. The maritime esplanade planned on the banks of the Ter has been extended as far as the embankment where the buildings sit. The racing boats built on the base, veritable machines comparable to a Formula 1 car, create quite a spectacle in the water and on land. The translucent buildings, lighted from inside at night, act as the city's lanterns, and are symbolic objects on the landscape.

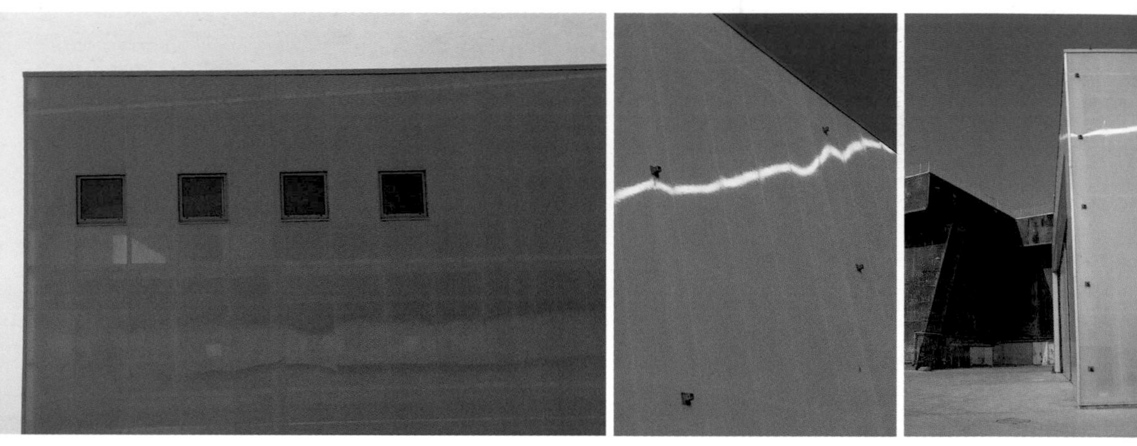

Oasis Apartments

Hans Peter Wörndl, Wolfgang Tschapeller, Max Rider

The Steinerstrasse, a project also known as the Oasis Apartments, consists of two housing blocks that contain a total of forty-eight dwellings in Salzburg, Austria. The appearance of this development, quite different from typical housing, is still marveled at and criticized by local residents and passersby. The facings of the T-shaped row of houses, in an architectural style that lies somewhere between industrial and domestic, makes the construction stand out from the otherwise classical urban landscape of the city.

The site is for all practical purposes a regular square. It parallels the Steinerstrasse, from which the project takes its name. Two long bays divide the lot into two equal rectangles, serving to evenly distribute the apartments, which have either one or two floors. This particular treatment of the space reflects the geometric concept of the project. The soil extracted in the early stages of the project was recycled, a value important to this region.

Architect: **Hans Peter Wörndl, Wolfgang Tschapeller, Max Rider**
Location: **Salzburg, Austria**
Surface area: **37,674 sq. feet**
Construction date: **2001**
Photography: **Paul Ott**

The first stage of the operation was excavation work in order to lower the site. This process established the main entrance on an upper level and created a series of sunken private parks for the residents. The resulting terrain allows private space for each dwelling and keeps the public zones on a different level. The stairways and hallways consequently have become semi-private spaces, and provide access throughout the building.

After laying the prestressed concrete slabs that serve as the building's foundation, a light framework was raised. On the second level, an elastic membrane was used to clad the building. This membrane provides continuity to the façade because it takes the form of a single flowing horizontal panel. The mooring points of the membrane, tensed in two directions to counteract wind pressures, make the casing resemble a sofa-like padding. The overall look of the structure is thus perceivable as a sort of furniture construction.

Location plan

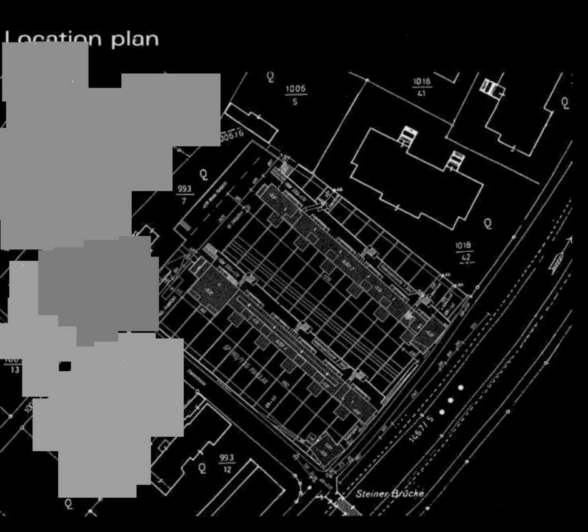

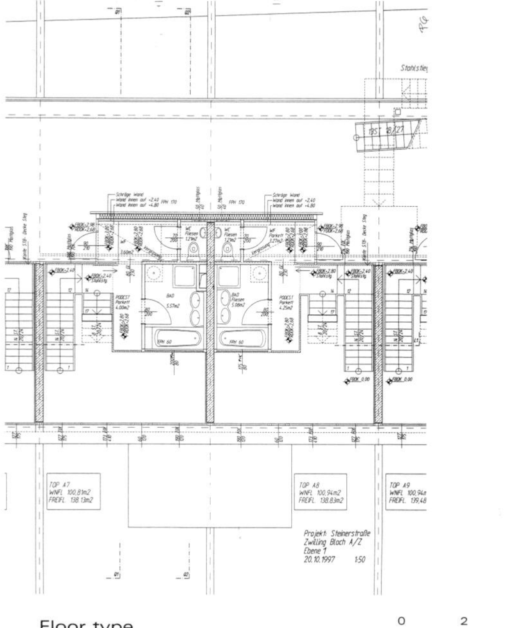

Floor type

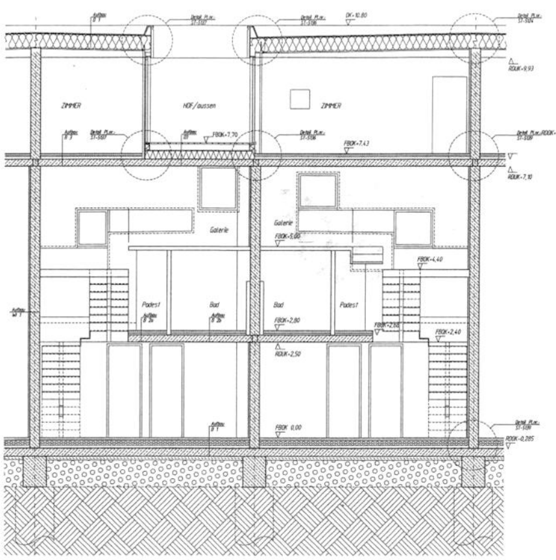

Longitudinal floor type

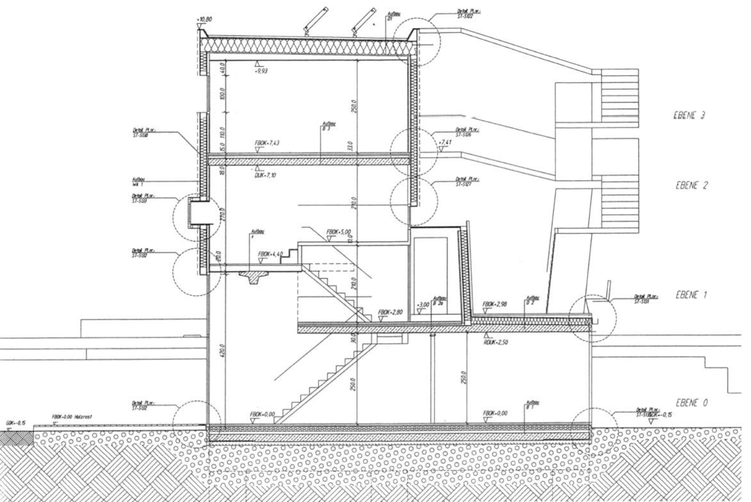

Transversal section

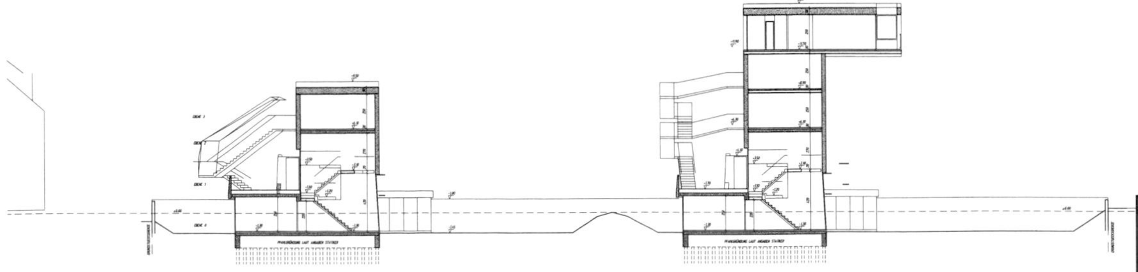

General transversal sections

The duplex apartment scheme includes exterior gardens and terraces. The finished product provides each unit with a character more like a separate house than an apartment in a series of clones.

Long wooden planters that run along the block outside the kitchen area of each apartment reinforce the relationship between the apartment and its garden. The kitchen and the rooms in the front of the house and those near the bathroom are totally finished in wood. The wood theme complements the outdoor planters that create a wooden side terrace, extending the interior space into the outdoors. A similar relationship with the exterior has been created for the upper level apartments, where the façade is set back at some points, generating other terraces which provide the same duality. In these cases, the terraces provide the best panoramic views available. Prefabrication of construction elements, such as the concrete and the elastic membrane that wraps the building, were decisions that aimed to make the project more cost-effective and to cut down on construction time.

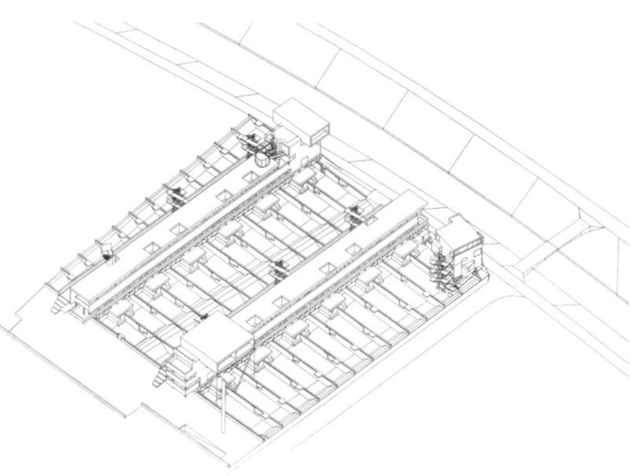

Aerial perspective

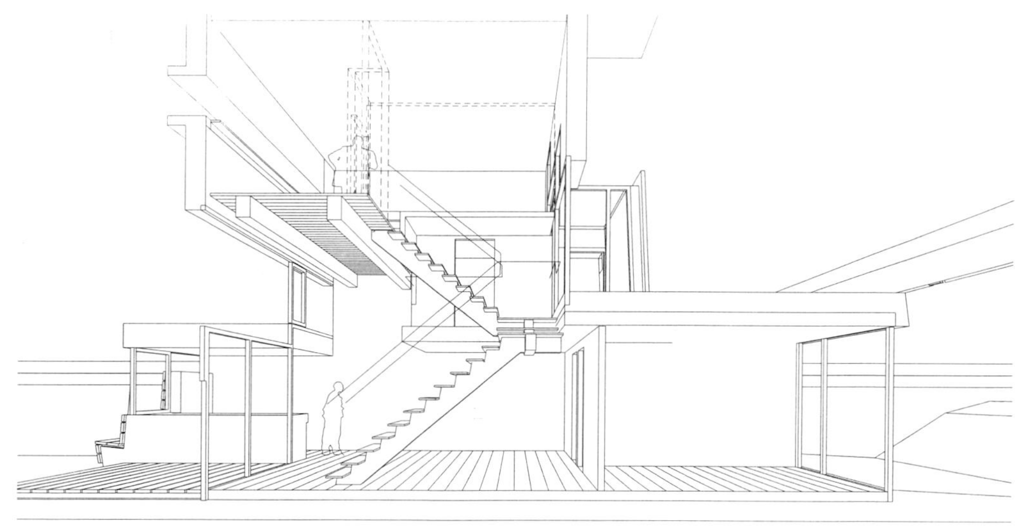

Section of a residence in perspective

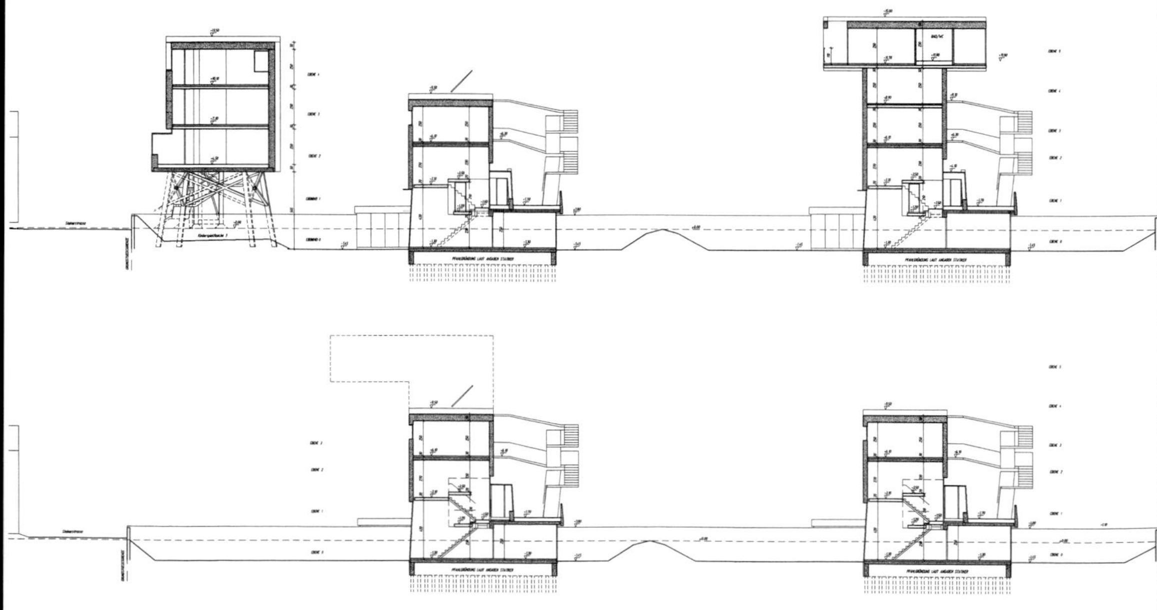

General sections

The varied mix of materials, such as wood, elastic membrane, and metal panels, endow every nook and cranny of the place with its own character. This tailor-made feel is underscored by the separation of the spaces and by the building's architectural style.

Doppelhofer

Wolfgang Feyferlik

This project's main goal was to create one building to house two different medical specialties. One part of the clinic was to be dedicated to alternative medicine, the other to general practice. The two practices would have to share some of the same facilities; however, they would also need to have their own individual space.

The whole design was put together in a steel construction that acts as the building's skeleton. The skeleton was then covered in different materials, primarily wood-derived panels. Other materials were also used, including glass and acrylic, both inside and outside the building.

The structure grew out of a steel frame thirteen feet wide anchored in place to a prestressed concrete foundation. This metal mount determined the general proportions of the building, but other bodies were added to the main mass. The walls are made of prefabricated wooden panels that measure a standard 12.8 by 7.2 feet (3.9 x 2.2 meters), attached to the metallic frame. The roof is comprised of waterproof pressed wood shingles attached to wooden beams supported by the main metal skeleton.

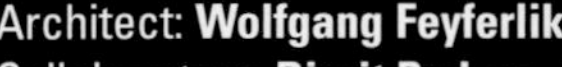

Architect: **Wolfgang Feyferlik**
Collaborators: **Birgit Rudacs, Fritz Moshammer**
Location: **Neudau, Austria**
Surface area: **2,300 sq. feet**
Construction date: **1995**
Photography: **Paul Ott**

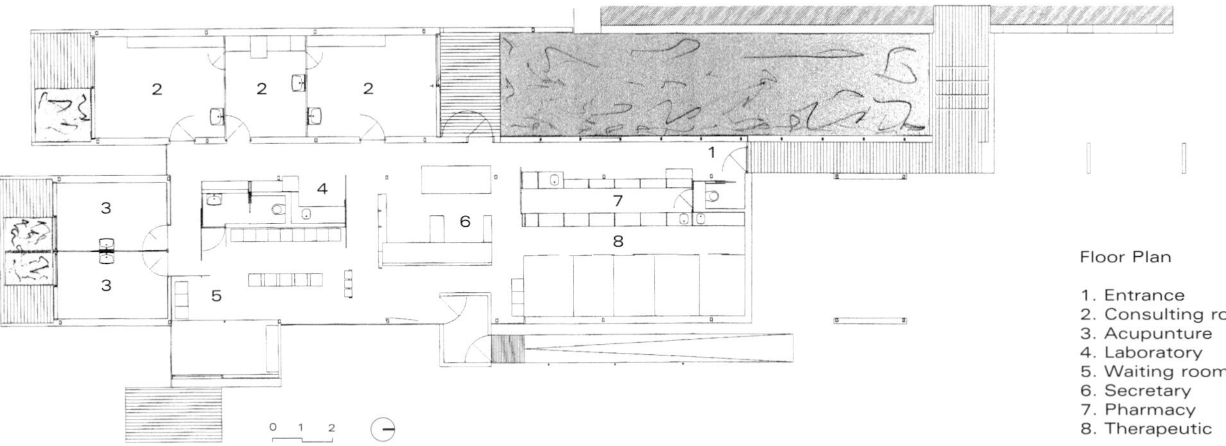

Floor Plan

1. Entrance
2. Consulting room
3. Acupunture
4. Laboratory
5. Waiting room
6. Secretary
7. Pharmacy
8. Therapeutic

The exterior employs different applications of wood: lath, wooden shutter, panels, and ornamental elements. This gives the construction a lightweight look and further connects it to nature. Inside, these same materials provide a sense of warmth and tranquility.

One of the goals of the clinic was to guarantee a quality private interior space as well as a sectioned-off administrative processing room. Taking advantage of the dense vegetation around the building, large windows were used to make the natural setting part of the interior. The natural element is therefore a key feature in the facings of the building and influences the therapeutic ambience.

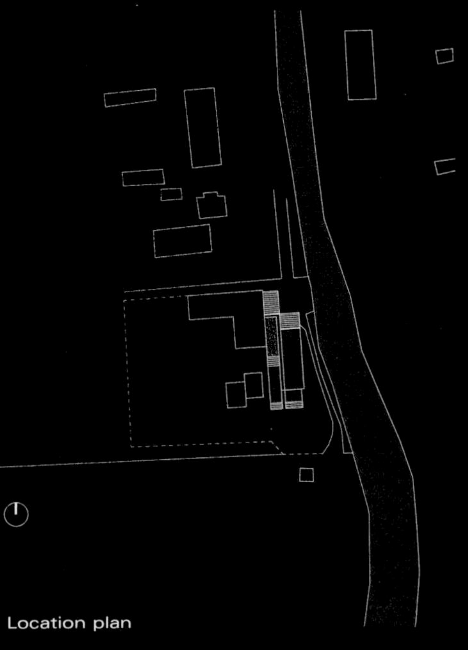

Location plan

The treatment rooms are grouped around a shared administration area. Each of the rooms is provided with its own open-air atrium, a small balcony with sliding doors and awnings (making the inside space appear larger than it really is). The sliding doors also make it possible to use the atriums as separate treatment rooms. These areas are ideal for such therapies as acupuncture or Chinese medicine, which require long periods of attention.

The interior of the center is a flowing space and can be divided to suit the treatment facilities. Lightweight wooden panels act as room separators, with colored glass and light shelves or bookcases. Natural light is abundant due to the use of oversized windows and skylights. These elements keep the natural light diffused throughout the building during the daytime and can, of course, be augmented by electrical lighting.

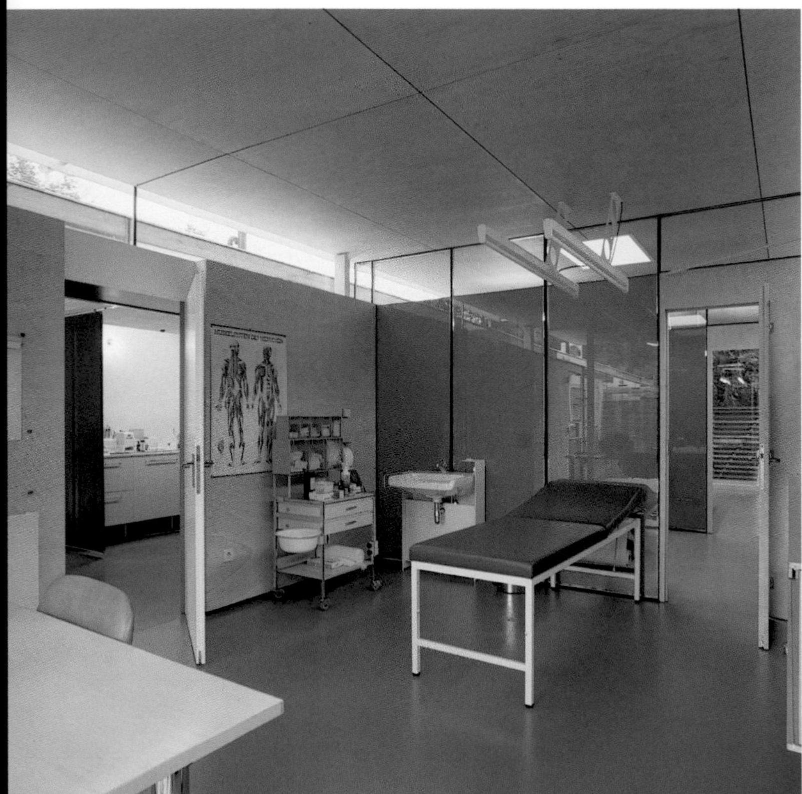

The large windows, interior patios, low walls, and skylights allow in abundant natural light and produce a soothing and pleasant atmosphere.

Spielfeld Border Control

Wolfgang Feyferlik

The public initiative to build a control center on the border between Austria and Slovenia resulted in a design competition held in 1995. Wolfgang Feyferlik won first prize and the commission to carry out the whole operation in which the previously existing buildings would be adapted and renovated, their roofs newly constructed and the outside spaces between buildings functionally connected. The project combined old, renovated, and new pieces into a single formal and functional whole.

The largest of the newly constructed parts, which serves as the connecting element in the set, is designed to house the administrative offices. A bridging structure more than 180 feet in length and made of steel has been placed between the northern and southern posts, which services arrivals and departures between the two countries. This building is supported at two points only: at one end on a concrete base slab laid on a previously existing building, and by a concrete column, at about two-thirds of the building's length. The construction was prefabricated in two parts and assembled on the site with the help of a crane.

Architect: **Wolfgang Feyferlik**
Location: **Spielfeld, Austria**
Surface area: **32,292 sq. feet**
Construction date: **2000**
Photography: **Paul Ott**

The bridge structure parallels the highways, seemingly denying its purpose as a border element by appearing more like a link between the two territories. The walls, floor, and roof were made of prefabricated wooden structure, and the façade is constructed of a tensed textile membrane that provides protection from the wind and the rain. Stretching this fabric slightly provides the western side of the building with protection from the sun.

The space below the structure is lit by lamps that are attached under the corrugated acrylic elements, thus creating a striking texture on the underside of the building.

The border control building is entered at either of the points that touch ground. A staircase paralleling the cement slabs and spiraling around the column leads to the upper story. The offices inside look west and are flanked by a corridor that looks out onto the eastern horizon.

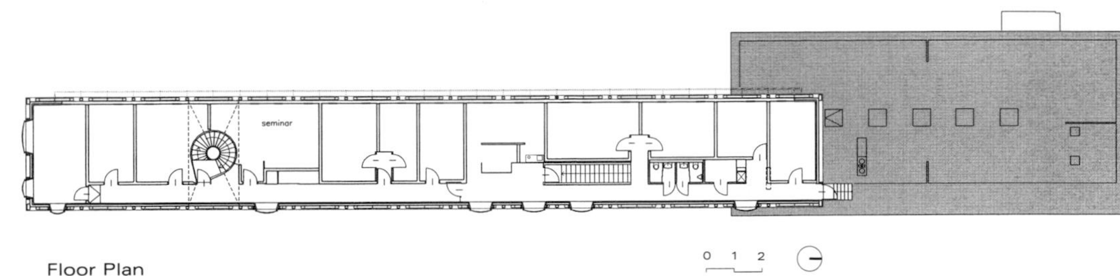

Floor Plan

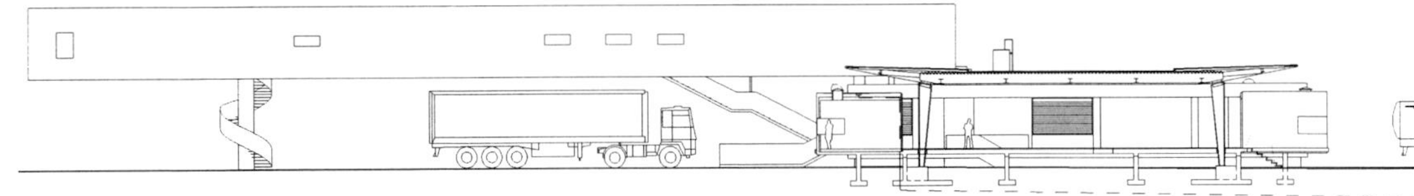

Longitudinal section

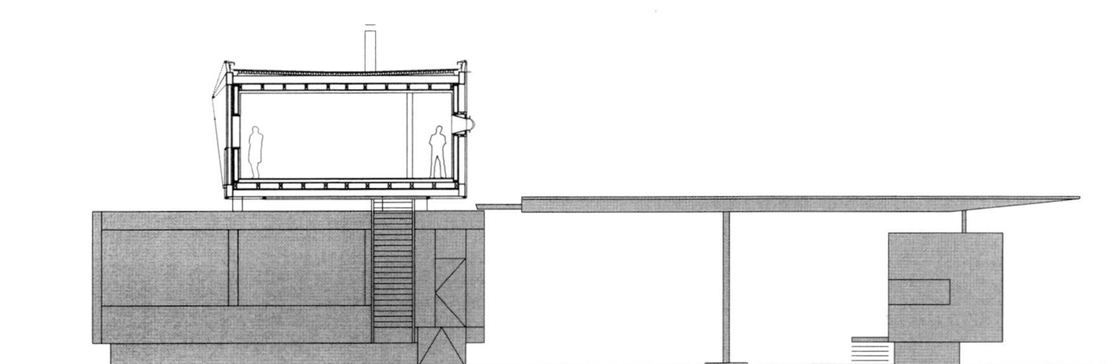

Transversal section

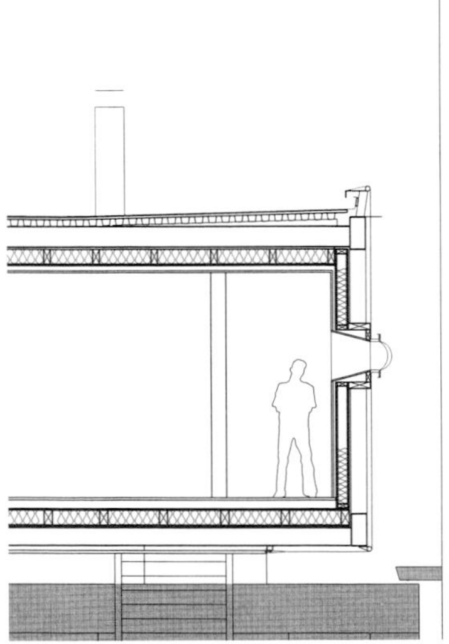

Interior detail

The prefabricated system took less on-site time to build by using simple, basic, low-cost materials, but each detail was carefully planned. The exterior ornamentation brings to mind the dome architecture used in the windows on the eastern front, with very small gauge metal screens used in the handrail of the stairs, the mooring of the external membrane, and corrugated acrylic in the false ceiling downstairs.

In contrast with the cold outer face, the interiors are very warm and refreshing. The finishings are made from wood veneer and glass dividers allow the natural light to permeate the area and make it appear larger.

The building's composition, proportion, and use of prefabricated materials can be read as a comment on the trucks and cars that pass through here on a daily basis. Spielfeld is like one more vehicle parked at the border.

Higashi – Osaka House

Waro Kishi

This house, although it appears to be a modern marvel, is not a heroic expression of contemporary technology and the latest industrial products. To the contrary, it was built by covering a steel skeleton with prefabricated cement panels, which were then fitted with large steel window frames.

Architect Waro Kishi wished to explore the contemporary potential between the industrial and the vernacular. Hence, the house is the complete antithesis of the monolithic. No question, here, of a one-off prestressed concrete edifice where all of the parts depend on a complex overall principle. The question is, instead, that of a type of frame that can be put up by anybody, anywhere in the world.

Taking the same approach that he had employed for other projects, such as the house in Nipponbashi, Waro Kishi used common local building methods and thus arrived at one given solution for this place, this site, and this environment.

Architect: **Waro Kishi**
Location: **Osaka, Japan**
Surface area: **2,002 sq. feet**
Date: **1997**
Photography: **Tomiko Hirai**

1K
マンション
東大
☎722-5530

The building is set in an Osaka suburb, in a residential district that has been fractionalized by various private railway lines. The property is very close to a commercial street that suffers severe gridlock and leads to a train station surrounded by condominiums lacking in urban character. This dwelling can fit in only two directions: one looks onto another house, and the other looks onto the street itself. These conditions are a constant in any Japanese metropolitan suburb.

The site's features and Kishi's desire to create an atmosphere that would not reflect the harsh conditions led to the construction of a project that opts to view itself instead of the environment. An entire interior world was created with one patio following after another, relating to the spaces of the house and simultaneously creating a series of other spaces that channel the walkways through the house or the outside terraces.

1. Entrance
2. Parking
3. Store
4. Patio

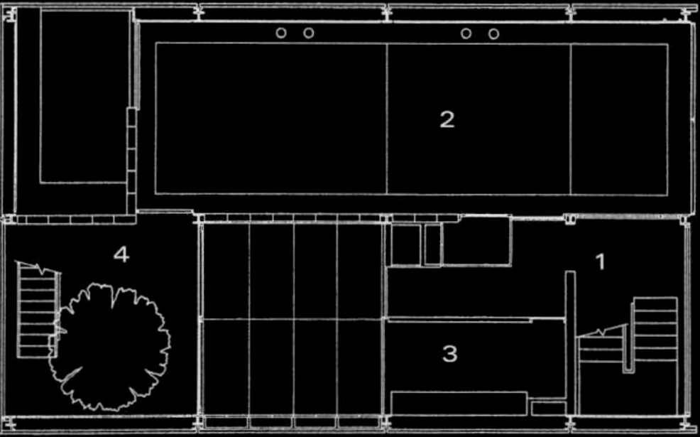

Floor plan

5. Bedroom
6. Terrace

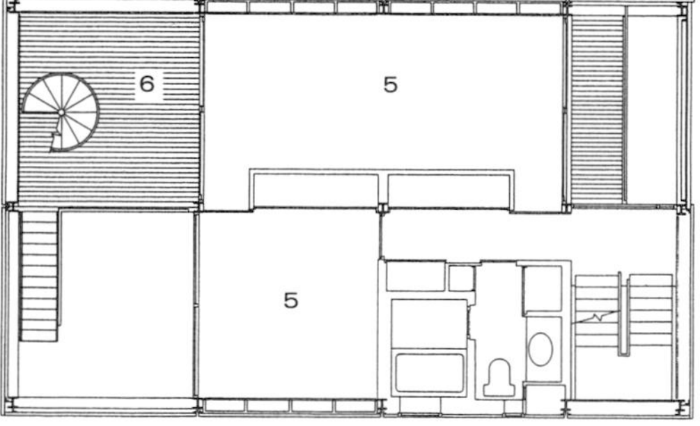

First floor

7. Living room
8. Dining room

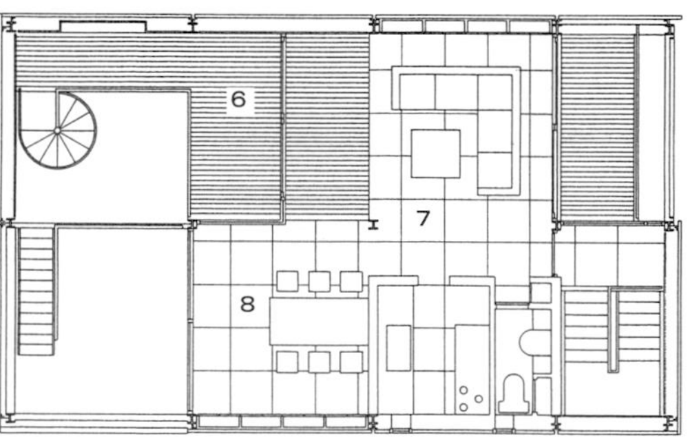

Second floor

0 1 2

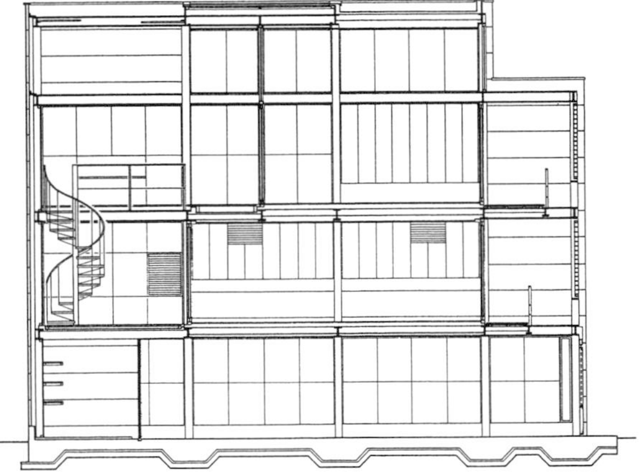

Section

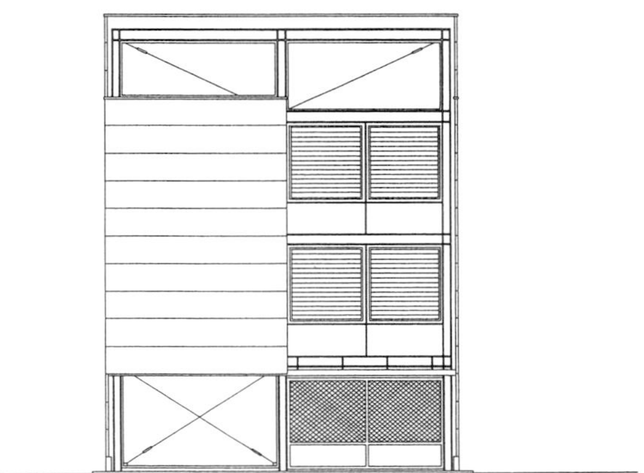

Front elevation

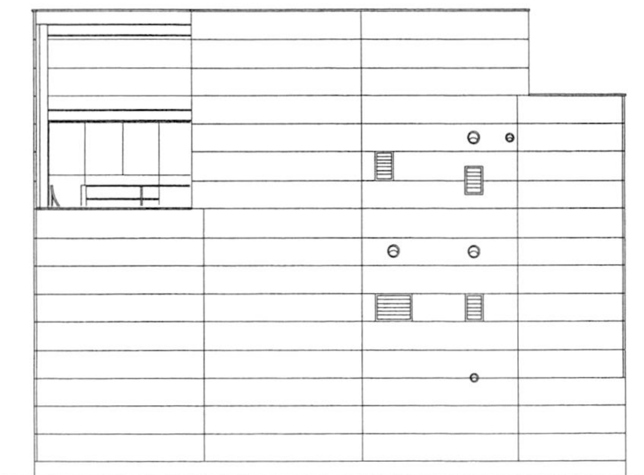

Side section

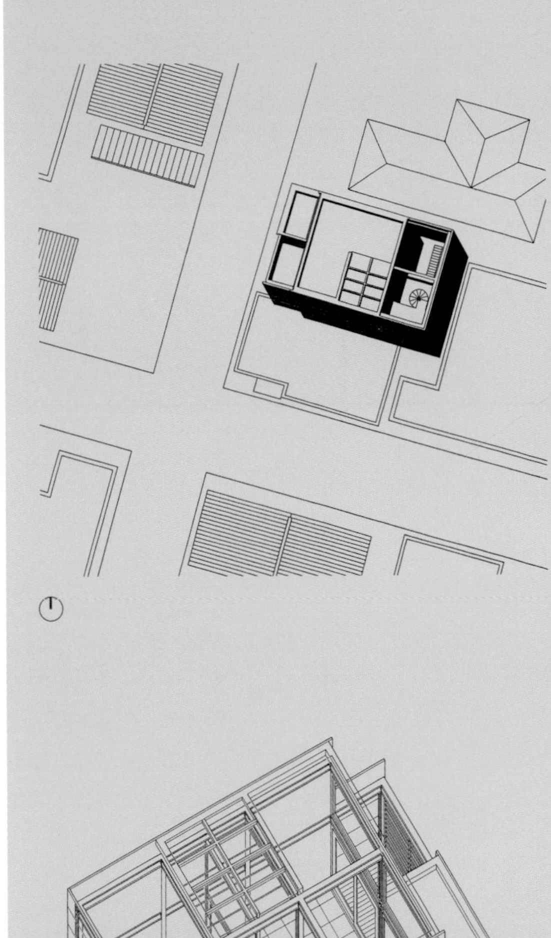

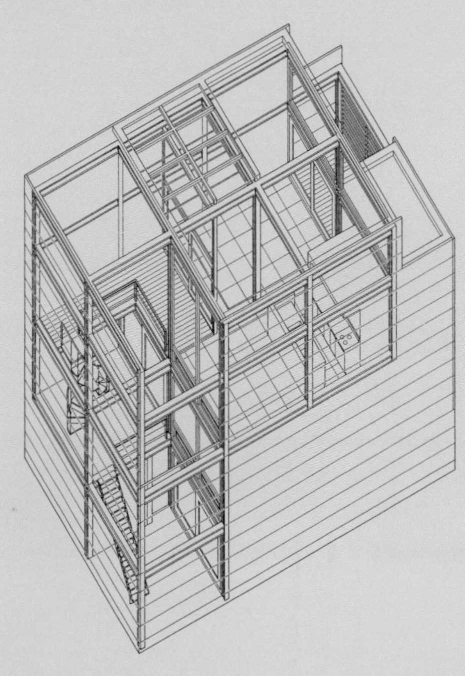

The large apertures, the glass, and the white surfaces of the structure, along with the interior spaces, go a long way in emphasizing all of the house's interior/exterior links.

The main structure is a façade with two divisions of 11.8 feet and a three-room depth of 11.15 feet. A stairway and a terrace occupy an additional space on the street front. Behind the building are a second stairway and a rear garden. This patio has a door to the top floor, which is the only door in the house that takes you in this particular direction.

Two routes serve to create a certain dynamic in the flow of the house, culminating in the living room and dining room (on the top floor). This space is 13.12 (4 meters) in height, and its roof is partially made up of glass panels that serve the purpose of orchestrating the transition between the exterior patio and the interior of the dwelling.

The main front is no more than a few yards off the terrain, but it has a character all its own. This is especially easy to appreciate with the shape of the building and its relationship to the neighborhood in which it is built. This difference in height produces a space that strikes one as removed from the city and at the same time open to it.

Bus Stop Michael Culpepper, Greg Tew

No matter where you go in the North American city of Palouse, Idaho, you'll find the landscape dotted with shelters that serve as bus stops, used mainly during the hard winter months. This bus stop shelter, designed for use in the town's neighborhoods, gave the architects an opportunity to experiment with architectural forms and methods while they solved the requirements for this specific site and function. The piece is of reduced dimensions and shaped by the double-width plywood panels used in its construction. Each 8 x 12 foot side comes from an identical doubled over sheet and acts as the load-bearing element of the structure, eliminating the need for the use of internal frames. The panels are made by gluing, bolting, and bracketing three layers of the quarter-inch panels together in a radial arrangement.

On the opposite corners of the shelter are eight braces holding the pieces together, which also support part of the load. The uppermost brackets hold the roof and ceiling in place. The construction was done off-site and, because of the techniques it uses, is closer to furniture-making than to large-scale architecture. Once assembled, the shelter was transported to its position and set up for its present use.

Architect: **Michael Culpepper, Greg Tew**
Location: **Idaho, U.S.A.**
Surface area: **64,6 sq. feet**
Construction date: **1999**
Photography: **Michael Culpepper, Greg Tew, Mark E. LaMoreaux**

Adaptable Modular Dismountable Light Mobile

The present site was chosen because of the magnificent southern views. The tiny lot on which this shelter stands is slightly above street level and protected by a thick stand of pines.

The braces supporting the seats inside the shelter run from front to back and are attached to columns set in the soil. This arrangement makes the curved shell appear to float above the site.

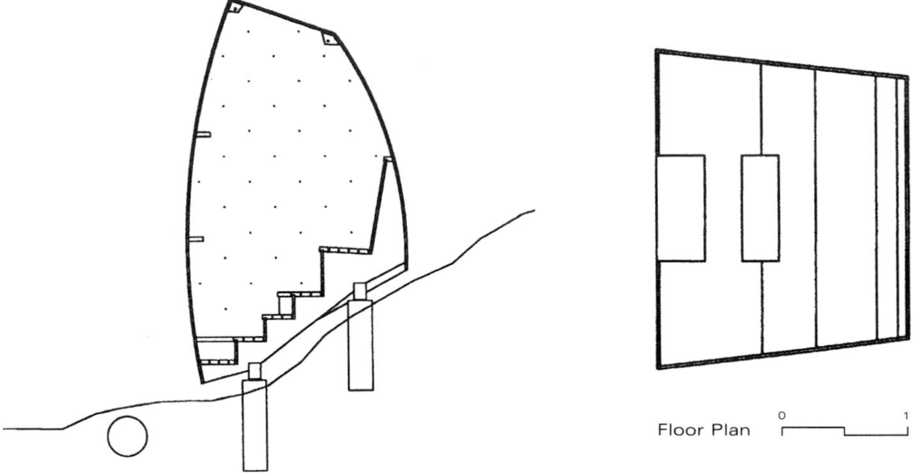

Section

The project is described by the architects as a public building designed to suit young travelers.

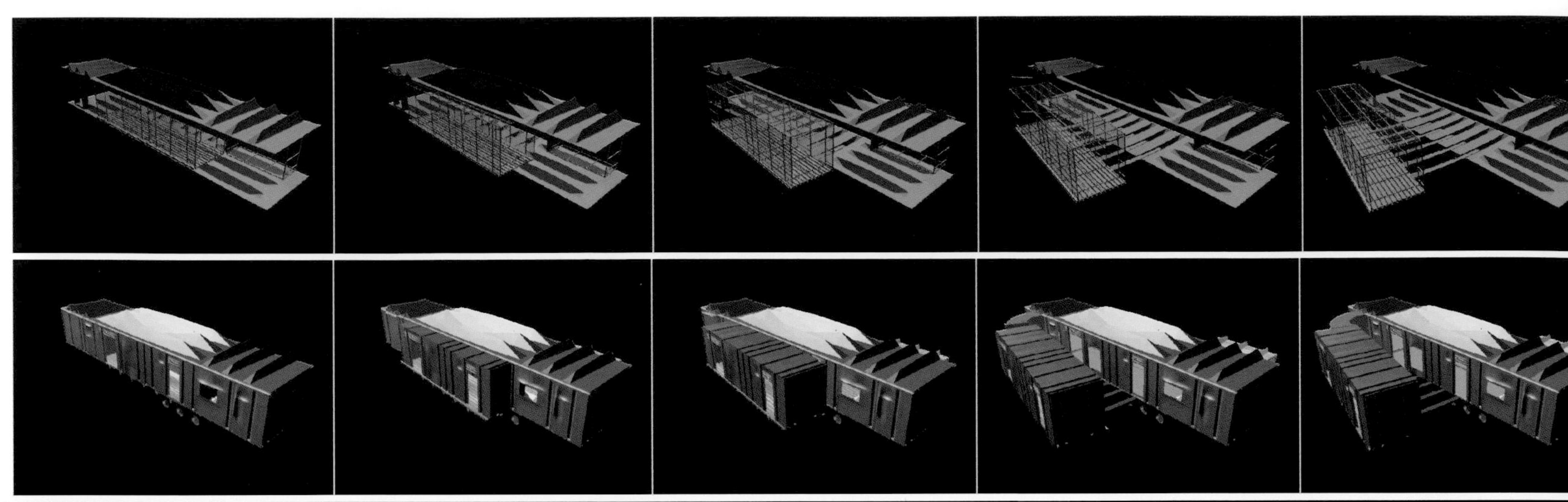

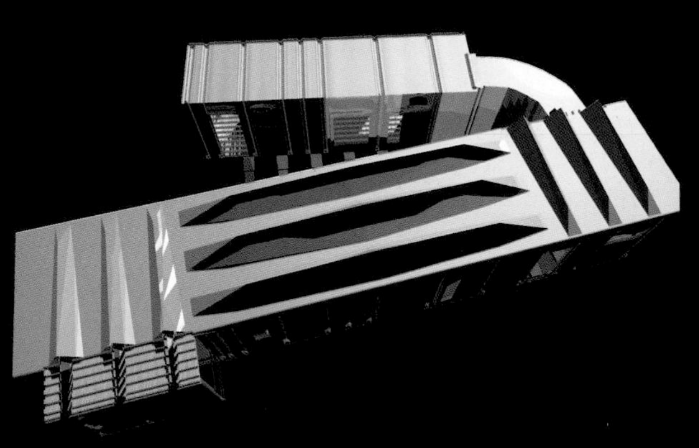

The new technologies make it possible to solve problems where the given characteristics of producing a series of homes can be made compatible with the building of individual private homes.

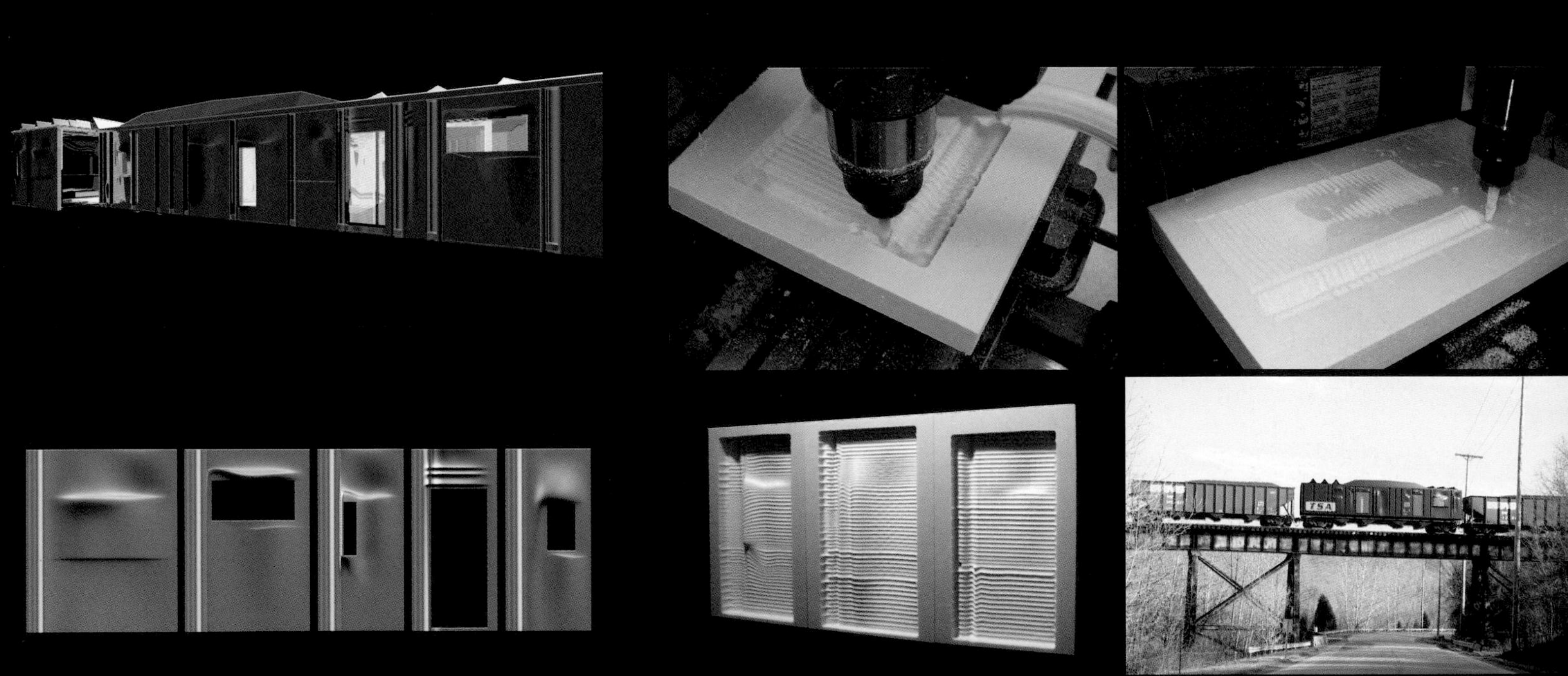

Architect: **Andrew Thurlow, Maia Small / TSA Architects**
Collaborators: **Jaimie Abel, Amanda Shadowens, Irina Verona.**
Location: **Mobile** Surface area: **646 sq. feet** Construction date: **2001**

Manufactured Housing

Through the use of computer-assisted production, new methods of manufacturing can bring about the creation of mass-produced building components. Nonstandard serial building systems, arising out of research into different kinds of new materials as well as new computer programs, create a new model where local variations make up a different structural composition. By means of this process, the architect can build an infinite number of private houses that are, for all practical purposes, identical.

Achieving production of a specific model within the bounds of this mass production includes a new look at a duality that has never left architectural thought: the production and assembly process as opposed to manual and conventional building practices. The architect's job, simply, is that of adjusting these two mechanisms to bring about a project that adapts itself to each need while still taking advantage of serialized production such as the reduction in costs and building time. This research tends to approach the transformation in production of three main elements: structure, decoration, and space.

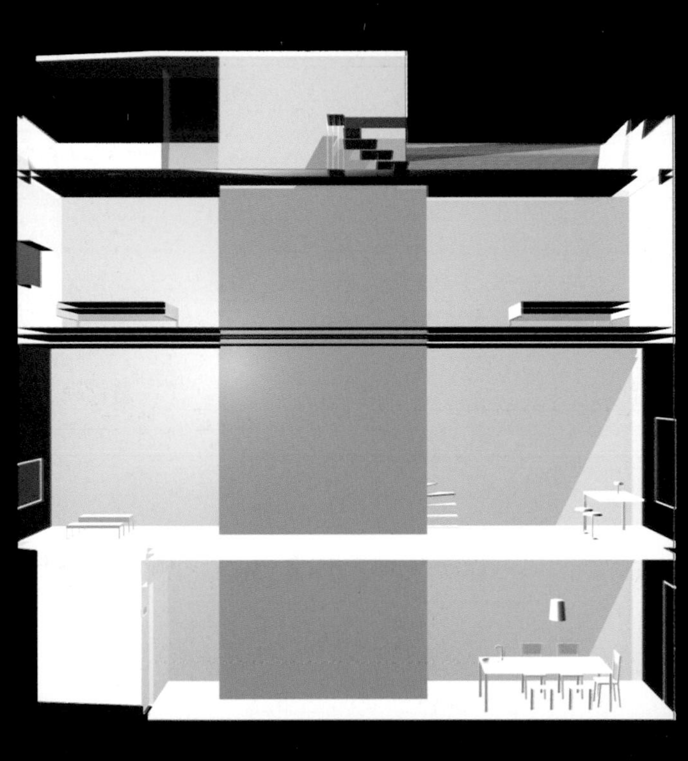

This contemporary house not only confronts problems of urban density, reflected in its compact interior spaces, it also meets the demands of quick and simple construction.

Architect: **Oskar Leo Kaufmann** Location: **Anywhere**
Surface area: **1,292 sq. feet** Construction date: **2000**

House in Mailand

The main idea behind this project is the creation of modular dwellings that are built from prefabricated wood elements. It is possible to group them together in a row inside a densely populated and congested city environment. Theoretically this could bring about greater living space and comfort and still be cost-effective and accessible to the average family. Confronted with the small areas available in city centers, the project uses different techniques to create a feeling of space and freedom, including different rooms, lighting impressions on each floor, a terrace, a rooftop swimming pool, and the plain and simple interior design. This is a building where production, transport, and assembly have to be fast and simple.

The design itself, based on simplicity and the urban character it reflects, is a three-story volume, mainly of wood, found in the floors (solid parquet), walls and ceilings (wooden panels of 3.9 feet). The outside shutters are blinds that give a greater or lesser degree of privacy. They also serve as ornamental features on the multi-layered façade. Inside, the panels are painted a cool cream tone to bring out the other interior decorating elements in earth tones and bright colors. The complete assembly of this striking house can be done in two or three days.

Pavilion of Yamaguchi Prefecture

Katsufumi Kubota

The building that serves as the pavilion for the temporary trade fair exhibition space in Ajisu, Japan has its work cut out for it. It has to take in more than 2,500,000 people in the course of a three-month show, which perhaps explains the plain, clean lines that back up the functions required of this public office building.

The main conditions determining the design were the physical characteristics of the site and the image which Yamaguchi Prefecture wished to project with this structure. Constructed on land reclaimed-from the ocean, the site looks out onto the natural richness of the blue water, the bright sky, and the wind. The aim was to start with these conditions and provide the visitor with a feeling of peace and vitality.

The composition of the space accommodates a program of basic requirements and simple activities: there is a theater, containing a gigantic screen 269 -square- feet wide and 129 -square- feet high, a lobby, an exhibition gallery, and a maintenance patio. The formal language, of the same simplicity as the program itself, manages to unify these functions to create a clear and uniform building.

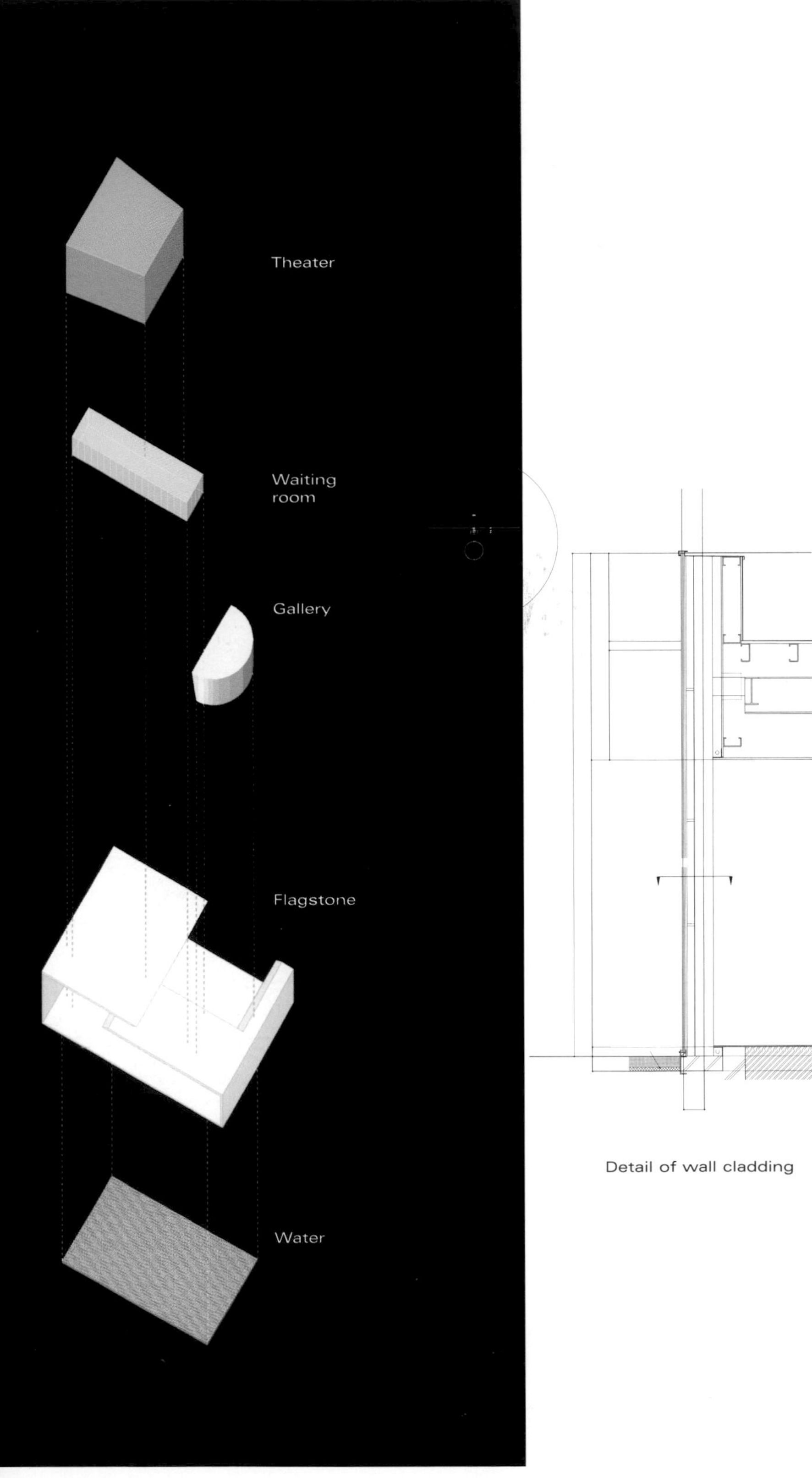

Detail of wall cladding

Architect: **Katsufumi Kubota / Kubota Architect Atelier**
Location: **Ajisu, Yamaguchi Prefecture, Japan**
Surface area: **18,300 sq. feet**
Date: **2001**
Photography: **Mitsuo Matsuoka**

The white canvas facilitates changing moods between different activities, whit an open corridor of flowing, adjacent spaces. This basic gesture molds a white space, minimal and highly abstract, which reveals its origins in a rather extraordinary way. The fluidity and the vibrant quality of each room is largely due to the proportion of the different dimensions.

The pavilion's design includes a system to dismantle the building piece by piece. Such a system had to be portable so that it could be transported to other site is and reconstructed and reused for new activities.

The challenge was to a building that was temporary and inexpensiveand that not only met the characteristics of structural stability against wind and seismic threats but also against snow and other more permanent elements.

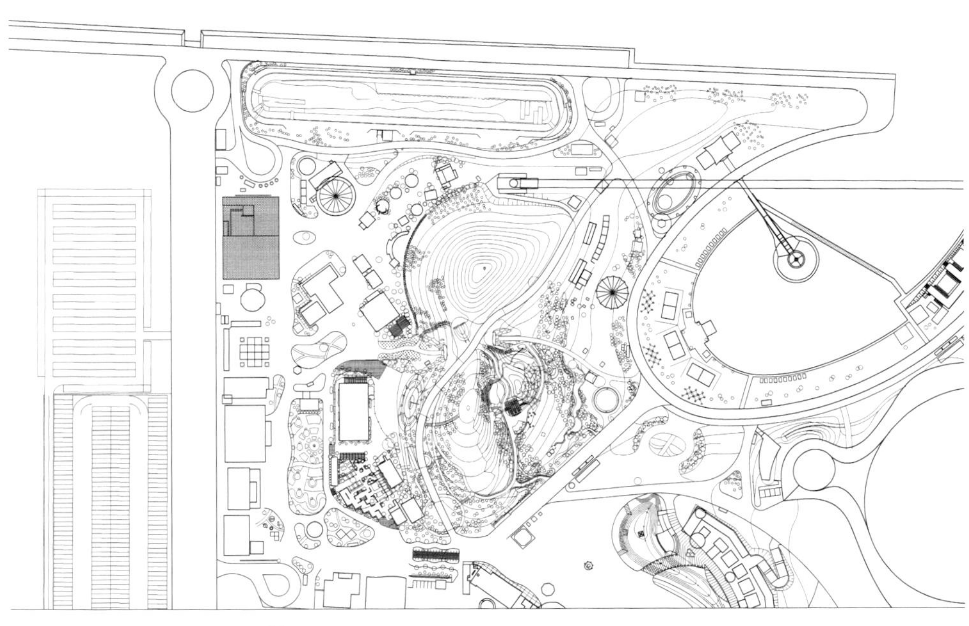

Location plan

The project's architect, Katsufumi Kubota, describes the pavilion as an opportunity for visitors to experience a stimulating visit centering on an atmosphere of freedom and relaxation.

Costs were cut by using economical industrial products that still allowed great formal expression. These materials include rough-textured cement, galvanized steel, and a plastic material for the facings. The aesthetic result was achieved by working through each detail with highly qualified personnel capable of doing artisan work.

The desing is organized around a large central patio, in this case a kind of reflecting pool with the rooms set around it. The entrance is a slender fissure between this space and the gallery, drawing the eye to the pool. From this lobby space, one may enter either the theater at one end or the exhibition space at the other. The curved wall of the gallery molds itself into an outside staircase leading to the terrace.

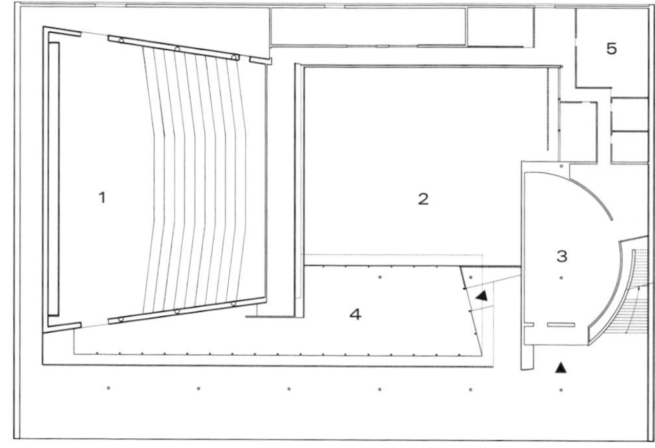

1. Theater
2. Water
3. Gallery
4. Waiting space
5. Patio

Roof plan

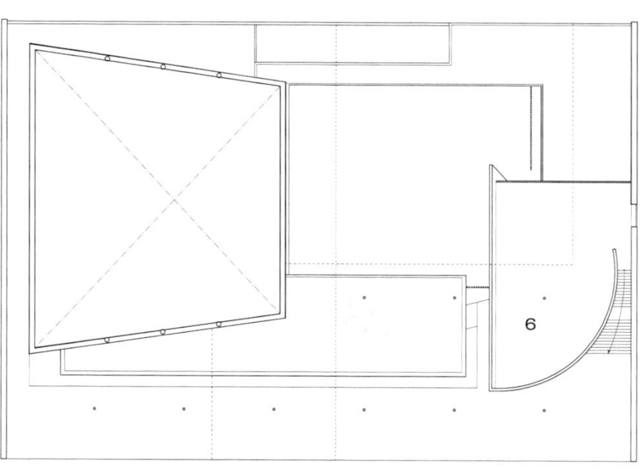

6. Terrace

Upper plan

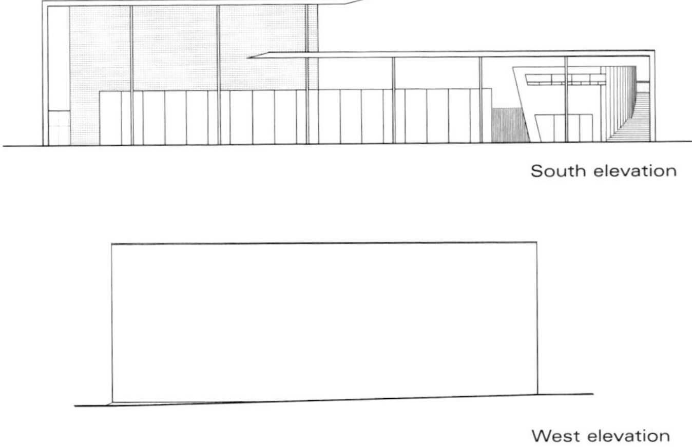
South elevation

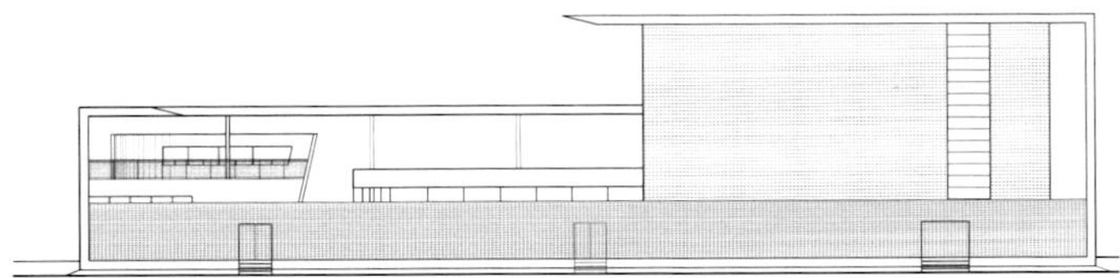
West elevation

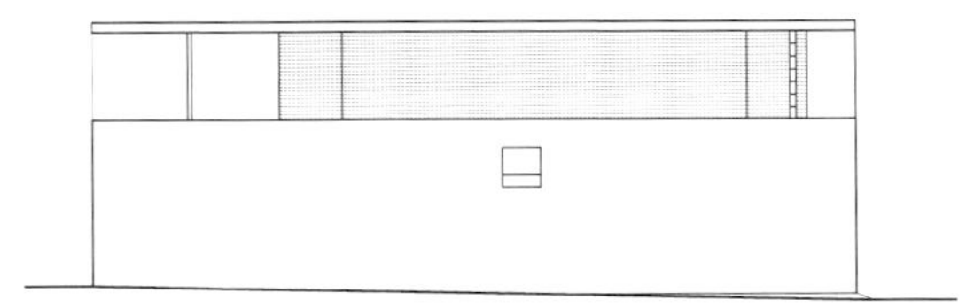
North elevation

East elevation

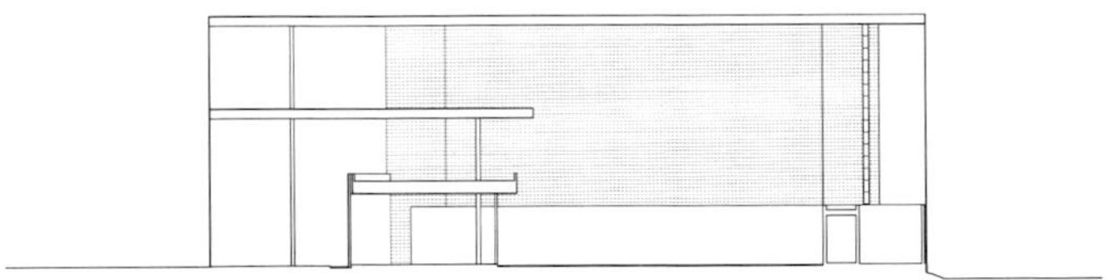
Transversal section

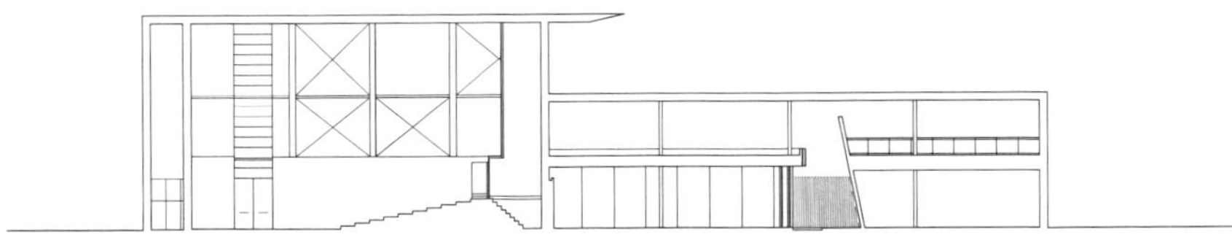
Longitudinal section

In spite of the elemental scheme and the use of the most basic materials, the building recreates a variety of spatial qualities. The folds of the concrete marquee running along the rooms yields intimate spaces that contrast the pavilion's details and its monumentality.

Expo'98 Kiosks

João Mendes Ribeiro + Pedro Brígida

This commission involved the creation for five different kiosk types whose design theme originated in the concept of multi-use features of the built structure as object or series of objects. This led to the production of a small building capable of taking on as many shapes as functional tasks. The construction generated a transformable object. It would eventually appear as a different image in each of its roles. As a pavilion for information, as a small store, as a warehouse, as a refuge etc. The object's plain, unadorned language-that of a wooden box, simply-adapts easily to a specific urban and natural setting.

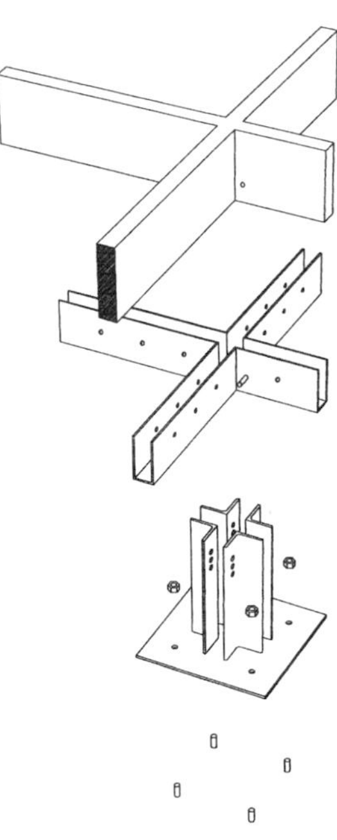

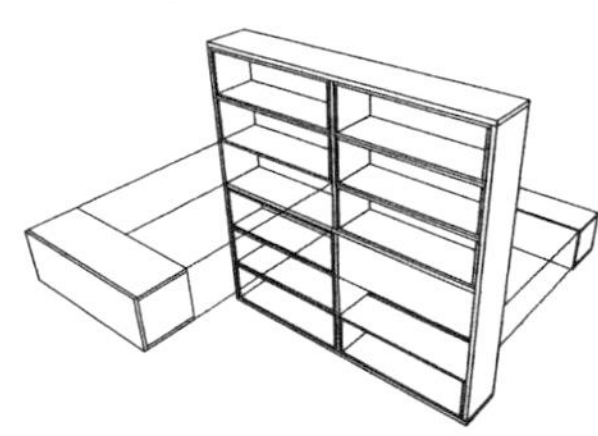

Assemblage details

Architect: **Joâo Mendes Ribeiro + Pedro Brígida**
Collaborators: **Cidália Silva, Edurado Mota, Manuela Nogueira, Nuno Barbosa, Susana Lobo**
Location: **Lisboa, Portugal**
Surface area: **250 sq. feet**
Date: **1998**
Photography: **Sérgio Mah, Joâo Mendes Ribeiro**

Adaptable Modular Dismountable Light Mobile

The construction makes the edifice able to be easily removed from the place it grows from, considering that it was originally designed for a temporary exhibit. The spatial structure—formal and built—turns into an easily dismantled thing even if this means losing its properties of resistance to intensive use, the use implied by the passage of thousands of people.

Structural soul: trabeation in wood, T-shaped. The structure thus brings about greater systemic rigidity. The wooden elements, tied together by metal linkers, support the exterior/interior finishes that have been used. These are plywood panels with a special veneer that is normally seen used in naval construction.

The apertures in the piece, framed in the same material as the facings, are arranged through the use of a hydraulic system that makes it possible to turn them to a horizontal position, thus creating a fin to protect from the rays of the sun. Close the fins and you turn the building into a closed box, a sculpture the size of an urban building.

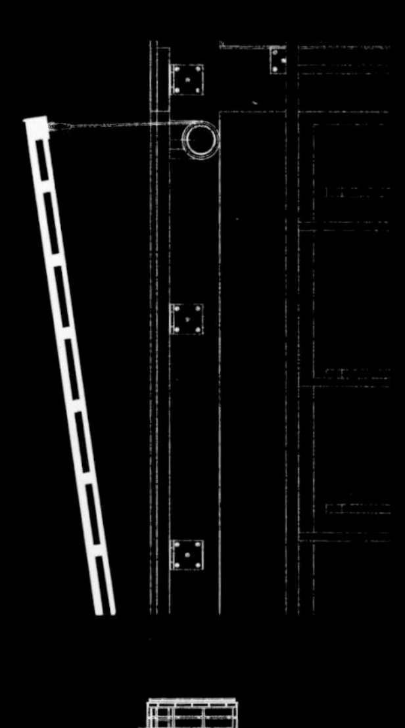

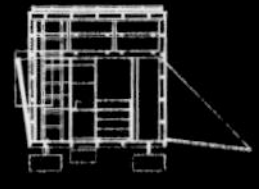

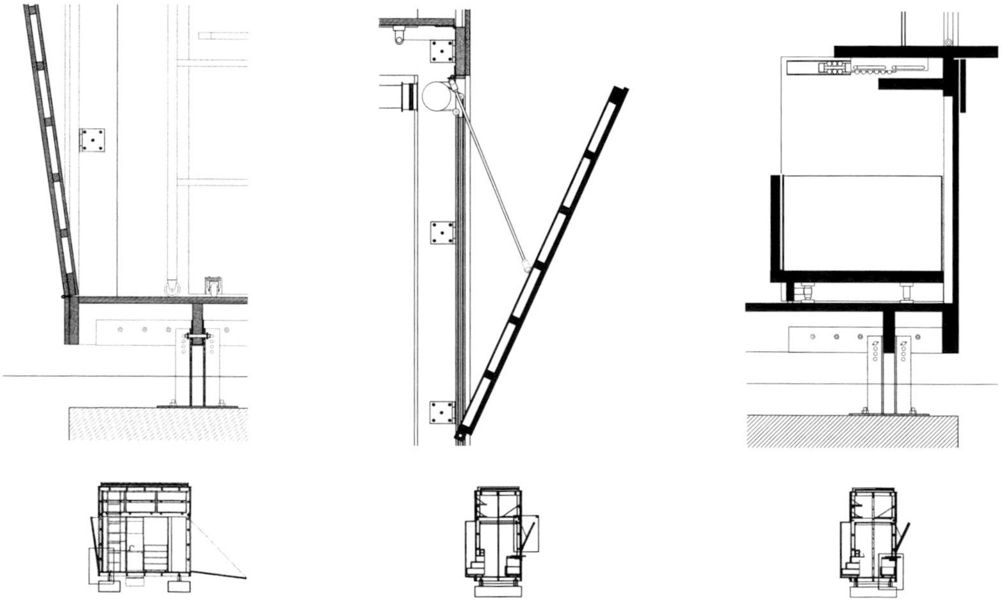

Details of folding doors and mechanisms

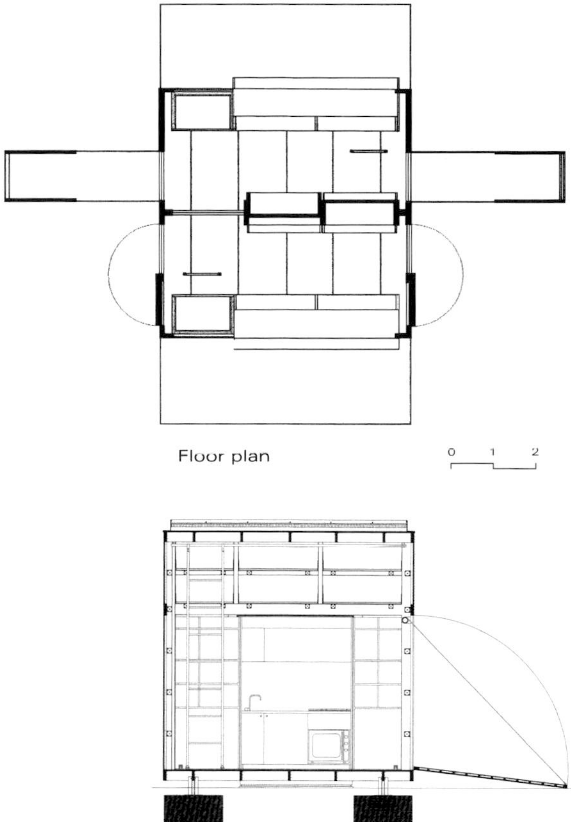

Floor plan

Longitudinal section

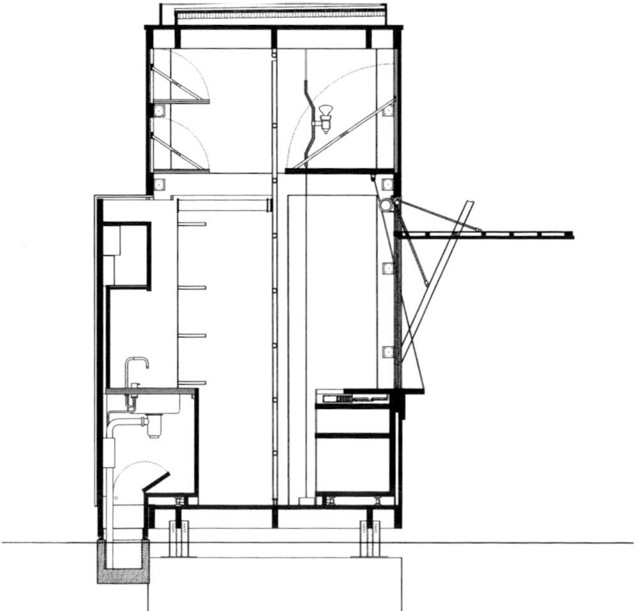

Transversal section

The mechanisms that have been incorporated and the different sizes of the folding doors yield a multitude of situations that adapt themselves to the different uses to which the structure can be put.

Servicenter

Steinmann & Schmid Architekten

This building is used as the center of general services for the Basle Trade Fair, in Basle, Switzerland. It is the first time the fairgrounds have had a single block that unites different services like offices, a press center, and an administration area. At the same time, its presence here works just like that of any other pavilion on the grounds: it will remain there for a period of time between three and five years, after which it will be dismantled and replaced by the construction of a high building conceived for this purpose. Since the building had to be done in a maximum of six months and would, at the same time, remain standing only for a few years, the architects and the clients decided to make Servicenter out of prefab pieces.

The building's pre-built parts are centered on the three main levels of the construction: the facings of the building, which also bear some of the weight; the frames and floors; and, finally, the roofs. The services and the press center are scattered along the building's three levels in different rooms that adjust to various needs. In addition to the wooden prefab panels, which include façade pieces and previously acquired windows, the stairs and the service rooms are also of prefab stuff. The assembly of the whole construction took something like five days, and the finishings something like five weeks.

Architect: **Steinmann & Schmid Architekten**
Collaborators: **Sophie Jaillard, Reto Zimmermann**
Location: **Basel, Switzerland**
Surface area: **13,347 sq. feet**
Construction date: **1996**
Photography: **Ruedi Walti**

Servicenter is located just in front of the Round-Patio Hall, one of the fairground's signature buildings, It was designed by architect Hans Hofmann in 1953, inside the central plaza of the grounds. Because it is indeed central, the building is easy for anyone passing by or any trade fair visitor to enter. Consumers go into the building through a transparent door that leads to a vestibule giving onto the outside grounds. The intersection of the two levels in this 19.68-foot high space, 6-meters makes it easy for the eye to take in the entire structure on the vertical.

Taking advantage of new advances in prefabricated materials like wood panels and outside doors and windows, the builders added a large staircase to the façade. The dark color of the panels joining these become imperceptible as you move past and the façade takes on the appearance of a continuum. It is a simple cube framed in a kind of belt with sides. The windows, also framed in this metal structure, incorporate folding as part of the natural ventilation system.

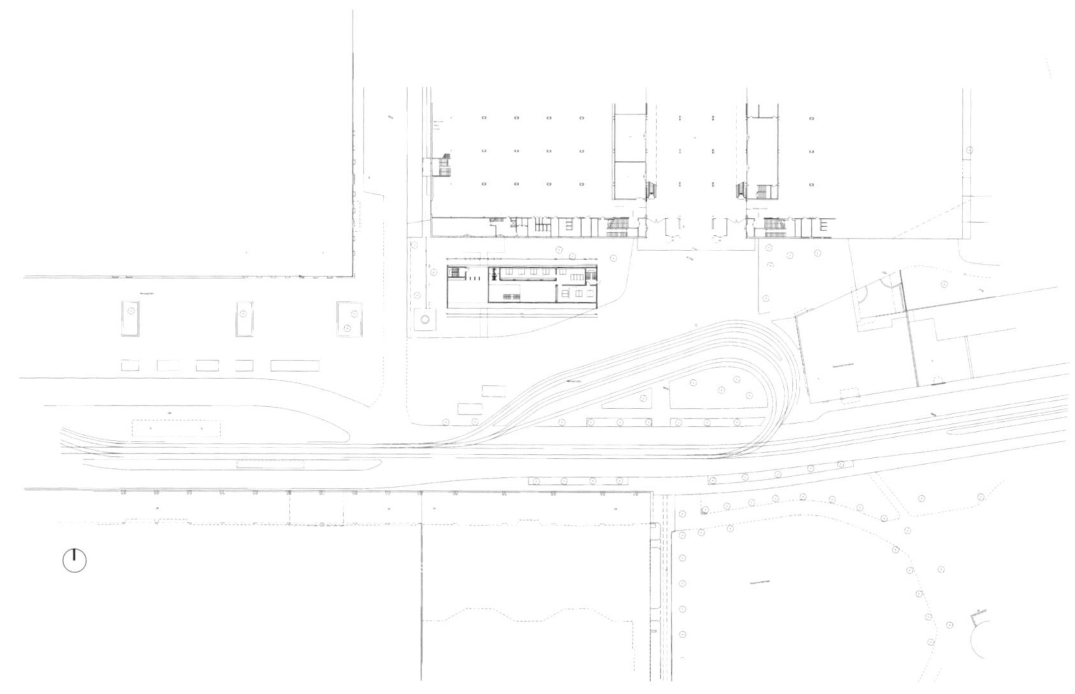

Location plan

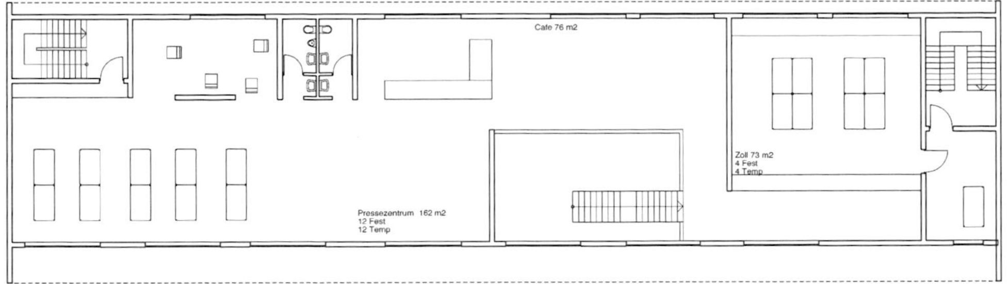

First floor

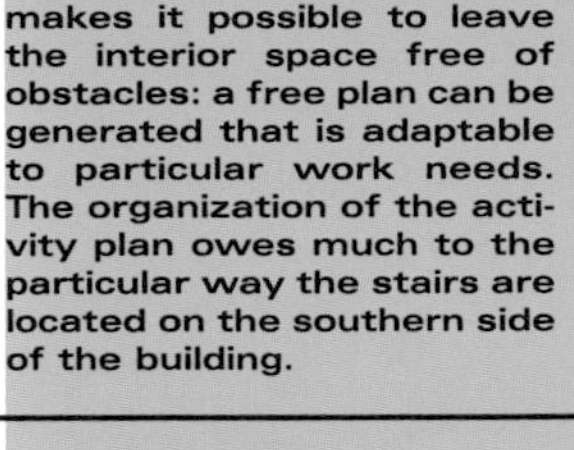

The structure's perimeter makes it possible to leave the interior space free of obstacles: a free plan can be generated that is adaptable to particular work needs. The organization of the activity plan owes much to the particular way the stairs are located on the southern side of the building.

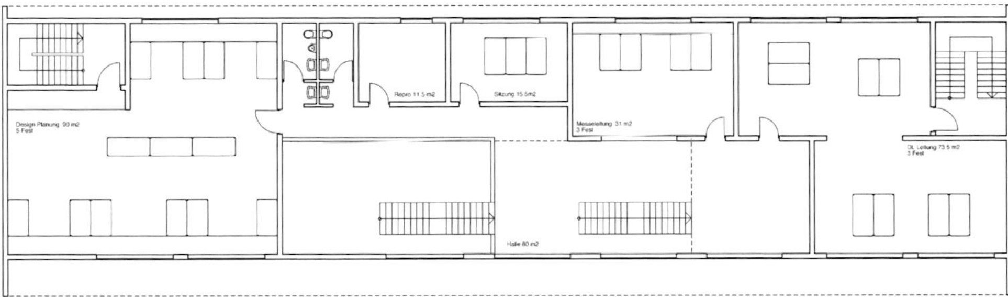

Second floor

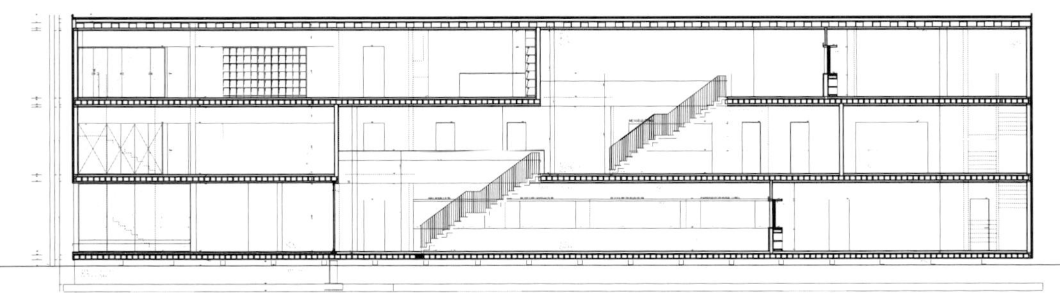

Longitudinal section

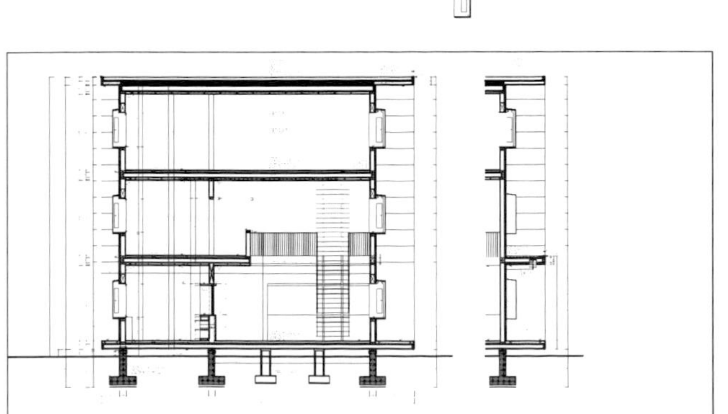

Transversal sections

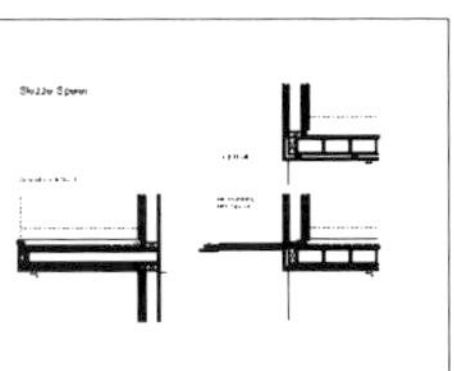

Details of junction between elements

The three floors, a total of more than 11,302 square feet of service area, interlink spatially through a large vestibule 19.68 feet in height that interfaces with all of the other spaces. In the building's interior, the work counters accessibe to the public. Moreover, visitors can use the well-marked route to find their way easily through the building.

The stairs, which are of lightweight construction, access the different areas of the building. On the first floor, the press center gives onto the quiet waiting room and lobby. The second level boasts a wide array of rooms that offer different communication services, meeting rooms, conference rooms, etc.

Servicenter interiors are decorated in dark blue panels and white walls. There is a presence of cubic elements, mostly different types of furniture, in intense colors. Some of the interior walls function as sliding doors, allowing different combinations of opened or closed environments according to changing needs. This alternating interplay of color provides an interesting slant on the work counters and the way the building has been constructed.

The interior lighting plays on the same concept and geometry as the rest of the project. Small boxes of fluorescent light are suspended from the false ceiling in two different directions.

GucklHupf Hans Peter Wörndl

The strange name of this small piece on the shores of Lake Mondsee, Austria, may help to explain its origin. On the one hand, it is a distortion of the name of a hill near the site, Guglhupf, and on the other hand it is a typical Austrian cake known as Gucklhupf. The make up of both of these names depends on the verbs *gucken*, which means looking at something from a special point of view, and *hüpfen*, referring to the action of jumping from a high place in order to get a better point of view.

GucklHupf arose out of the architect's proposal for the Festival of the Regions, held in 1993. This event celebrates all of the cultural, artistic, and architectural achievements in the north of Austria. During the summer months, a whole series of exhibits, workshops, and social gatherings takes place there, with musical performances that develop the theme of "the strange." With this concept as a point of departure, the architect decided to use a small family property to create his own interpretation of the Festival of the Regions.

The design of the pavilion underwent changes during its construction. The architect, with the help of his collaborators, finished the essential part, however, to provide the main definition. The idea was not to reach a final phase in the work but to create a live object that would remain in a permanent state of change, even after its assembly had been completed.

Architect: **Hans Peter Wörndl**
Location: **Mondsee, Austria**
Surface area: **516 sq. feet**
Date: **1993**
Photography: **Paul Ott**

The principal idea underpinning the project was to create an inherent tension between opposing poles. The strange confronting the familiar, quietude beside movement, settling opposed to traveling, the sensation of refuge in relation to that of being away from one's home-these are the themes that are present in this box of multiple perceptions. It is reflected in the different ways that visitors approach the building in the summer months when it is accessible to the public. It has been used as a space for contemplation, for sitting and looking at the landscape, as the scene of parties, small gatherings, or readings. During the winter months, for private use, it has been used as a pavilion for bathing in the lake, for a small lodge, as a meeting hall, or to store nautical equipment.

The construction and the materials used are the same as for the manufacture of boats, applicable to the building in terms of its proportion and its exposure to the harsh weather conditions. The main elements are the thin sandwich panels of two-ply waterproof plywood. Between the layers is an insulating material nearly an inch thick that permits conditioning of the interior as the temperature changes. As additional protection the wood is coated with six layers of special transparent paint, also used in boat manufacturing.

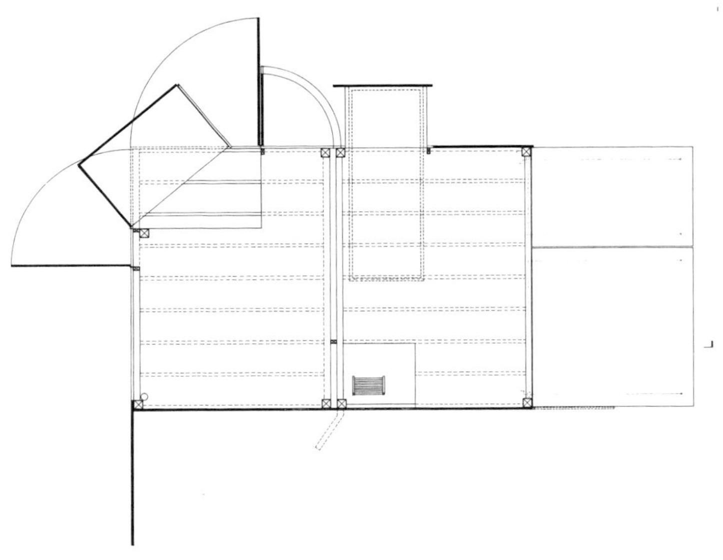

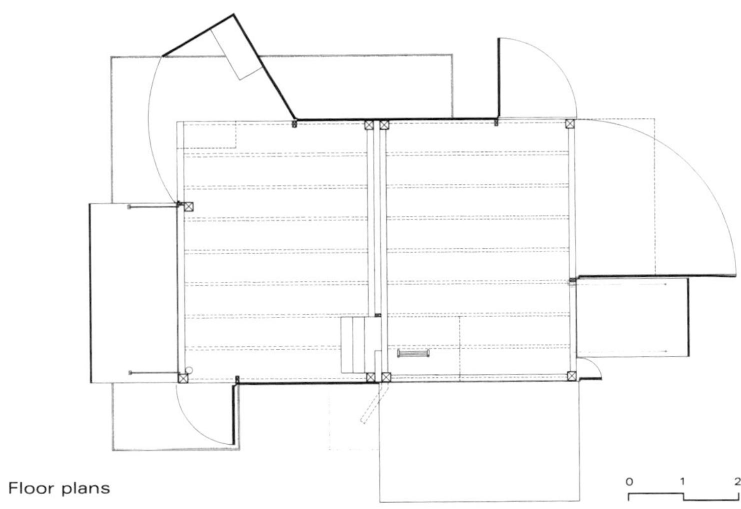

Floor plans

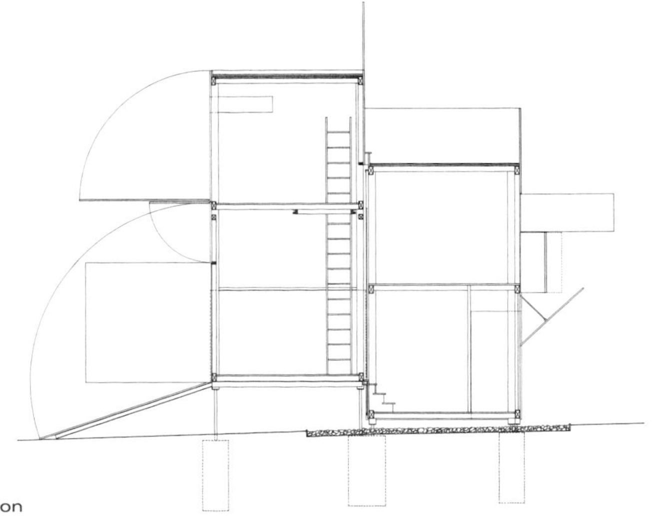

Section

GucklHupf can be interpreted in different ways, depending on the position of its folding panels. A closed container, a sculptural element on the lake shore, a pavilion, a kiosk in a garden, or a small house are just a few of the possibilities. The different arrangement of the elements emphasizes the relation between interior and exterior.

The interior is made up of a single triple-height room that can be subdivided into two, one above the other. The upper space gives onto the roof, which has been fitted out as a terrace. From here, one gets a panoramic view of the valley and the lake.

Although at first the local authorities made it clear that GucklHupf would remain on the site after the festival, as a sculptural element and exhibition space, the inhabitants of the region worked to revoke this decision. The pavilion was designated as a building that would have to be dismantled, according to the laws in force in the region. After searching for alternatives to keep it in place or to move it, GucklHupf was finally dismantled and stored. It is currently awaiting another opportunity to be reassembled on another site to renew the dialogue.

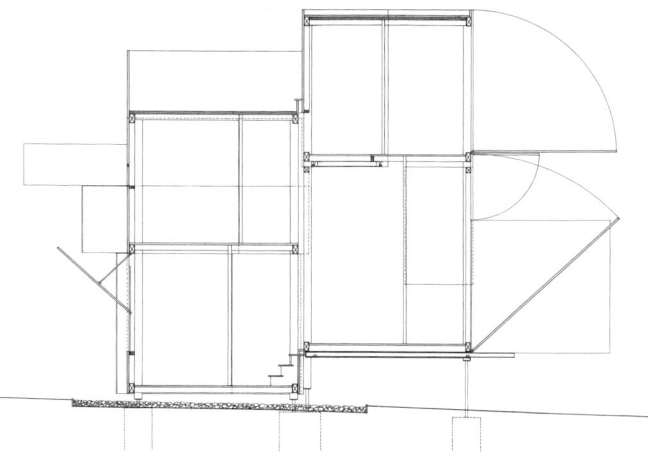

Section

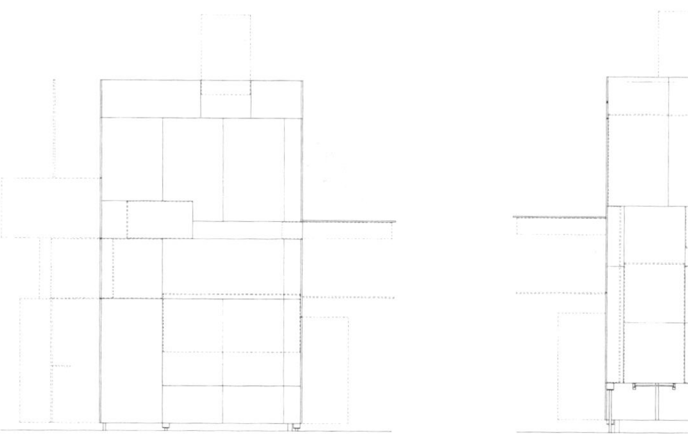

Elevations

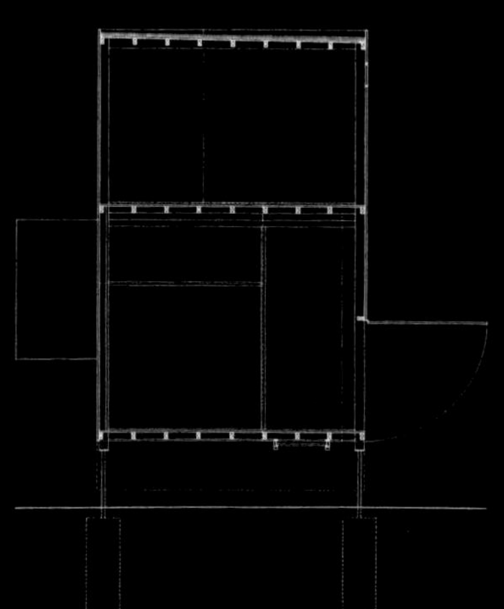

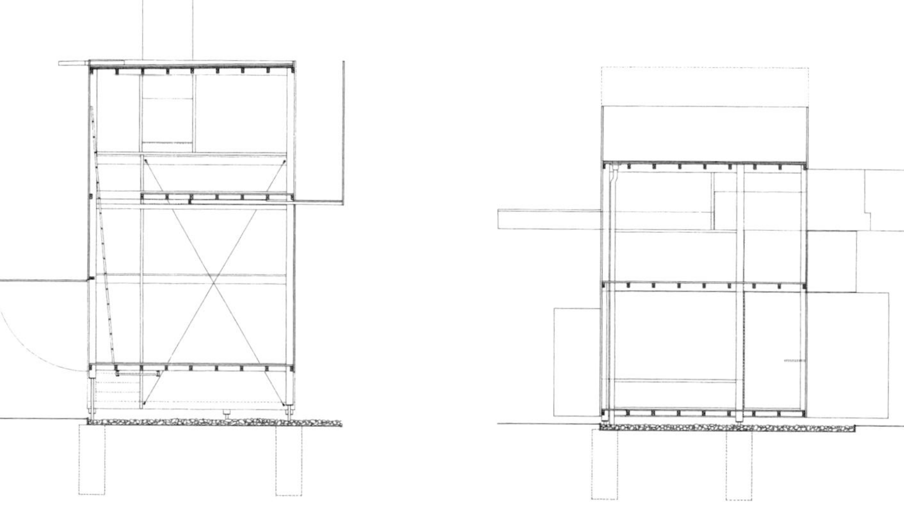

Sections

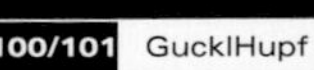

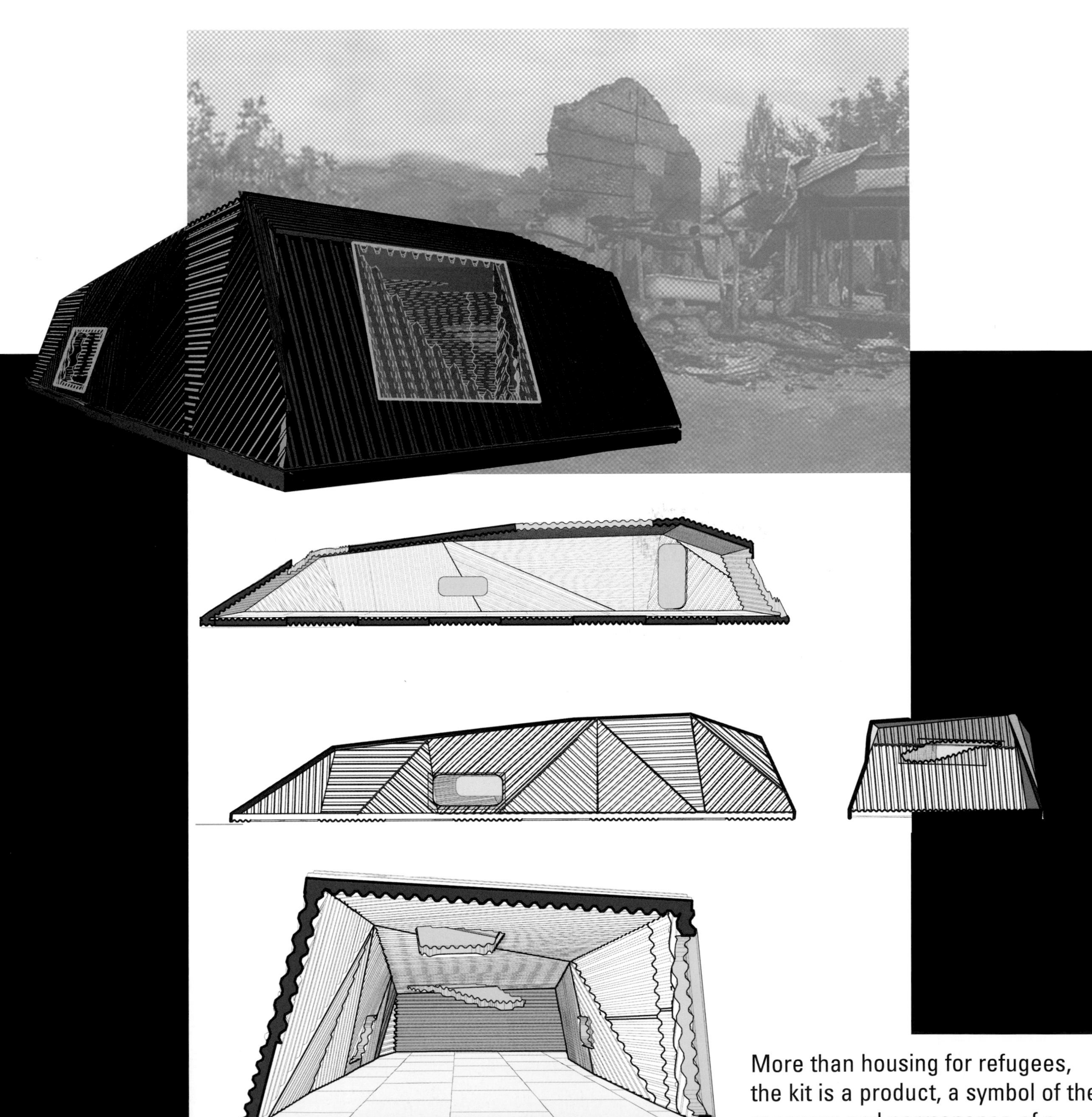

More than housing for refugees, the kit is a product, a symbol of the recovery and permanence of a people.

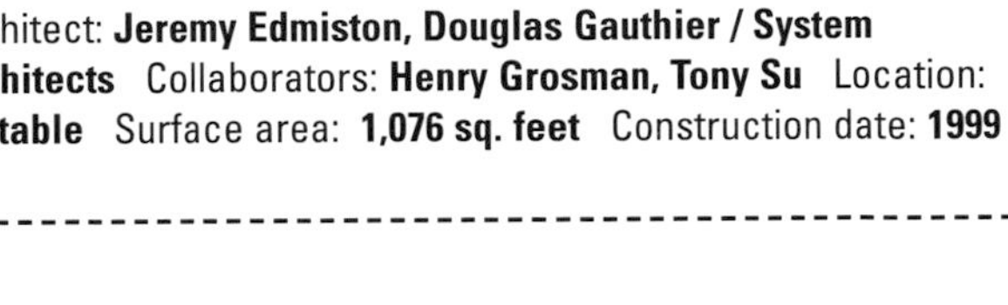
Architect: **Jeremy Edmiston, Douglas Gauthier / System Architects** Collaborators: **Henry Grosman, Tony Su** Location: **Portable** Surface area: **1,076 sq. feet** Construction date: **1999**

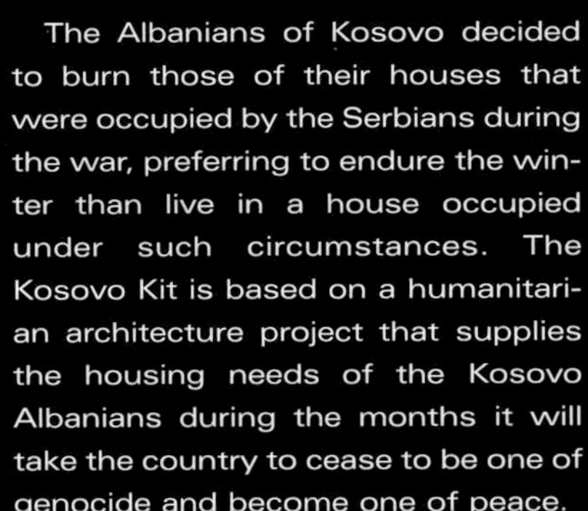

Kosovo Kit

The Albanians of Kosovo decided to burn those of their houses that were occupied by the Serbians during the war, preferring to endure the winter than live in a house occupied under such circumstances. The Kosovo Kit is based on a humanitarian architecture project that supplies the housing needs of the Kosovo Albanians during the months it will take the country to cease to be one of genocide and become one of peace.

Here, architecture is not simply a passive past for a political action but rather something that affects the way in which the negotiation of a new reality is undertaken. The edifice/product becomes a standard in the battle for survival and reconstruction. Once society has again begun to dwell here, there will be a clear intention, not only political but also social, to reach bilateral agreements. The immediate environment, the domestic environment, will then become the first point of reference for peace, playing a fundamental role at the moment of constituting a state.

One of the advantages of the Kosovo Kit is the use of the latest building materials and techniques developed to create a refuge that can be transported, easily put up, and maintained at low cost. It is comprised of double panels of corrugated metal with an inner polystyrene core, creating a resistant element, both insulating and load-bearing. The panels are machine-cut very precisely according to computer-assisted design patterns. Each kit, which comes in a (12m x 3m x 2.5m) 472.4 in. x 118 in. x 98.4 in. box for easy transport, contains all the necessary panels to erect the refuge. The kit also includes the adhesive used in its construction and the transparent polycarbonate material for windows. The structure can be put up in less than a day, either in a camp or on the roof of a damaged building.

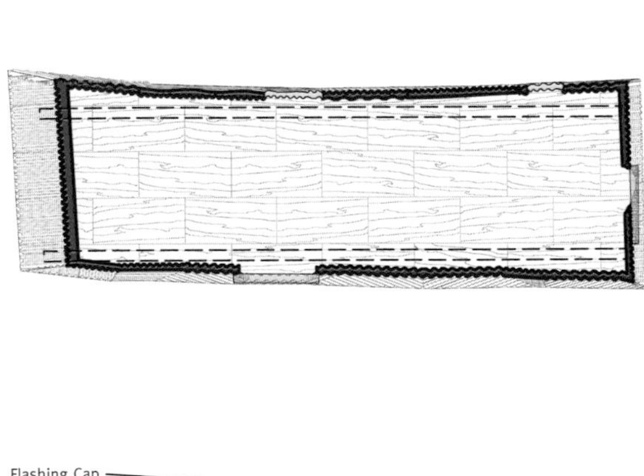

A system of rings, cables, and light frames are set up to create a small dwelling in the forest, with little or no damage to the tree that supports it.

Architect: **Softroom Architects** Location: **Mobile**
Surface area: **3,229 sq. feet** Construction date: **1998**

Tree House

This small, unique object is part of a series of projects by Softroom designed for the magazine *Wallpaper*. The theme can be summed up in one simple word, tree house, the old friend of our youth. The goal was to make a structure that could be attached to the trunk of a tree and cause minimal minimum impact to it.

The structure, a set of rings that supports a triangular frame, is designed to support different housing components such as a picnic table, a bath basin, or a folding bed. Not unlike other projects with similar typologies by these same architects, also done as commissions for publishers, the idea was developed over a very short course of time and allowed them to take some liberties in regard to the skeleton and the way the piece was put together.

It was an imaginative exercise and, to date, there are no models of this tree house that have actually been built. But with the right resources there is no reason why this model or one very like it should not be made. The project serves as a pretext to point out the relationship these architects understand to exist between landscape, architectural object, and interior.

Burghalde Alioth Langlotz Stalder Buol

Burghalde's design divides into equal size pieces that are presented in a repeated series. Additionally, the block of four semi-detached houses counts on an annexed piece on the lower part of the site intended to put up occasional guests.

Putting the building on the upper part of the terrain provided each house with the most magnificent panoramic views obtainable from this perspective. The block-of-four, entirely constructed of prefab parts, clearly alludes to the same kind of wooden crates we use for storage. The doors and windows are sliding types in the same material, setting in relief this aspect of the building while playing on the theme of carving out hollows in a solid.

The main framing device uses light-weight metal sets embedded in concrete slabs. From this simple point, all of Burghalde's elements, the doors, walls, façades, doors, windows, were put together off site. Transportation to the construction site and assembly then required a minimum of time. The system reduces building costs and assembly time for series reproduction. And at the same time it generates a homogeneous end product in the form of homes.

Architect: **Adriana Stalder & Leo Buol / Alioth Langlotz Stalder Buol**
Collaborators: **Thomas Henz, Philipp Esch**
Location: **Liestal, Switzerland**
Surface area: **7,320 sq. feet**
Construction date: **1998**
Photography: **Leo Buol**

Each house in the block is a module that repeats the same interior layout. There is a parking area beside the block. Space is available for a bath or, alternatively, another bedroom, studio or a small living room. The space beyond the foyer contains a staircase that leads both to a lower and an upper level. On the ground floor are the bedrooms, another bathroom, and a utility room where a washing machine could be installed. From here, a second stairway leads outside to a garden shared by the residents of the building. The upper level contains the living room, dining room, and kitchen all in a single large room.

The handling of the wood, the framing elements and the finishes demonstrates the way the project varies its style between the traditional and the contemporary. The interiors employ wood on the floors, veneer panels in the finishes, and solid wood staircases. The austere lines of the design in conjunction with the careful management of the materials make for an easygoing, neutral space that is very warm.

1. Entrance
2. Parking
3. Bedroom/Studio
4. Bedroom
5. Cellar/Laundry
6. Living room/Dining room
7. Kitchen
8. Balcony

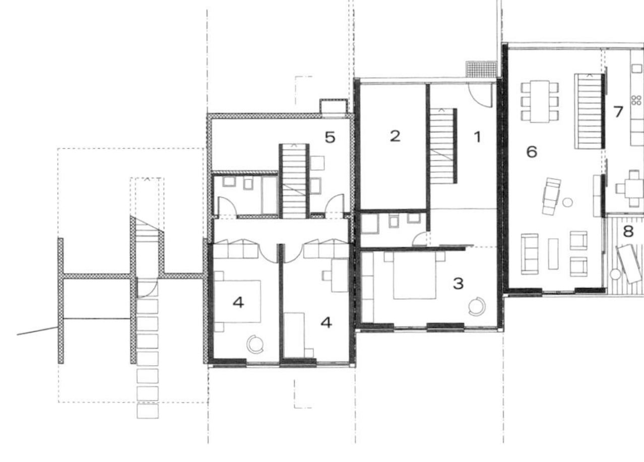

Floor plan

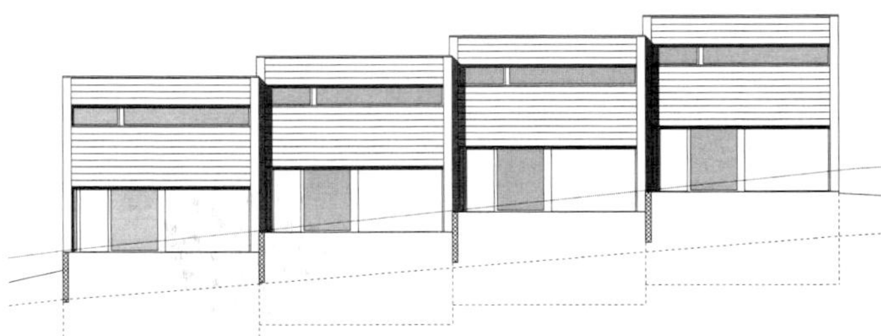

Interior elevation

Back elevation

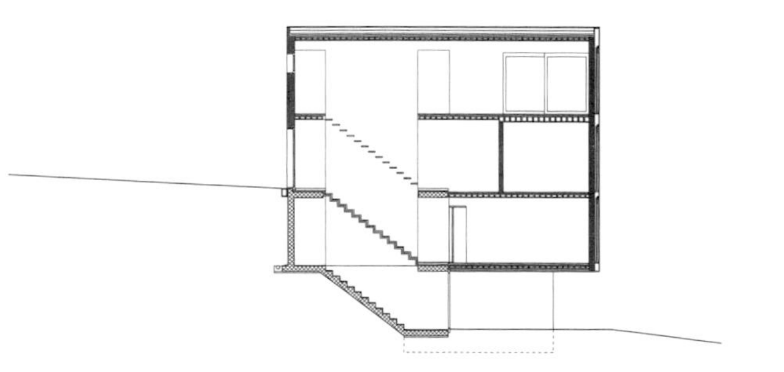

Transversal section

Pliezhausen Kindergarten

D'Inka, Scheible + Partner Freie Architekten

The design premises of this building-which contains rooms that are used for kindergarten activities-start from its location in an urban environment without excessively taxing the previously existing architectural and natural setting. The use of a basic approach and elemental framing devices has empowered a set of relationships linking the outer works and the building's interior. The children's premises are on the northern part of the site, in an orchard, and this is one example of the way that respect for the original features has been maintained to the greatest extent possible.

The building breaks down into structures of different sizes and varied arrangements that create a play among the open and closed, large and small bays. In toto, this choice interacts with the rest of the city's rich fabric by way of the terrain's simple interlacing paths. One of these, on the boundary, is notably marked by a wooden trellis used in viticulture.

So as to interrupt only minimally the cultivated areas and at the same time make a place for the light construction, the project was developed out of a system of wooden frames that compose the structural skeleton.

Architect: **Gabriele D'Inka + Albrecht Scheible**
Collaborator: **Hannes Streitenberger**
Location: **Pliezhausen, Germany**
Surface area: **6,458 sq. feet**
Date: **1998**
Photography: **Roland Halbe**

On rectangular plans, three small bays are arranged along the long axis of the set in an east-west direction. This simple straight line, however, is double height so as to accommodate the circulation for which the whole project was designed. All of the other parts of the building are accessible from here-both interior and exterior.

The three main bodies of the complex open onto the southern façade to exploit the panoramic views of the orchard and to gain the best conditions available in terms of the sunlight. The window frames are metal, providing a contrast with the wooden elements in the building itself and with the deck.

These wooden decks, one on each of the three added bodies, extend from the main foyer to the orchard. The children's supervision is managed from the foyers and the main classrooms (including special care units) with interlinking access between zones.

The general plan and the formal language of the kindergarten arise out of the idea of series construction with simple materials and prefabricated identical units repeated throughout. Wood is the element that provides the main ecological image, providing a cost-effective management at the same time.

General plan

0 2 4

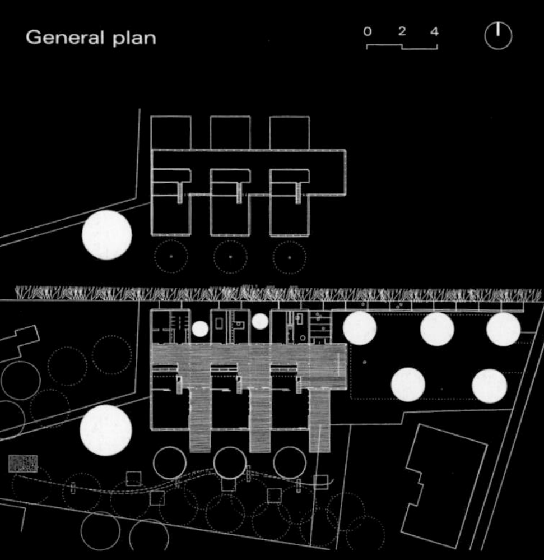

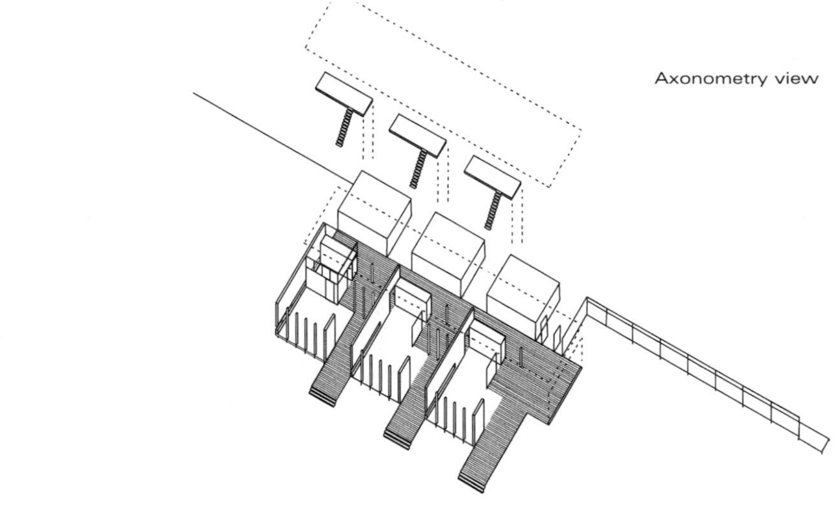

Axonometry view

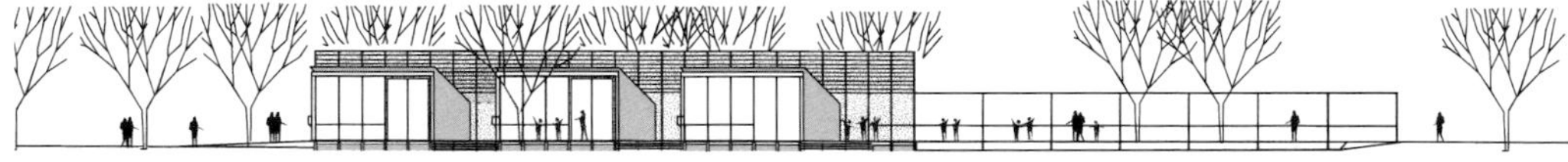

South elevation

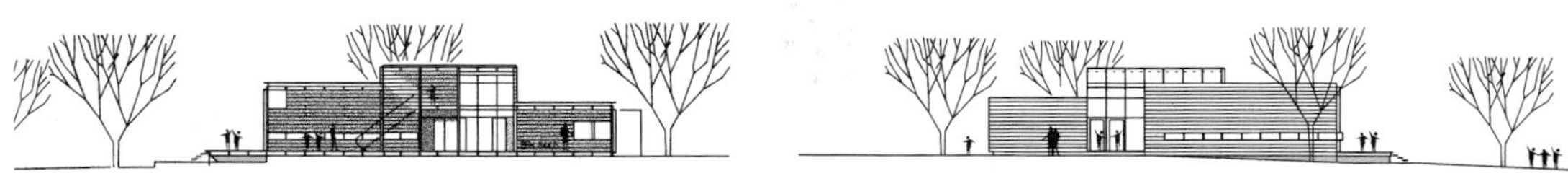

East elevation

West elevation

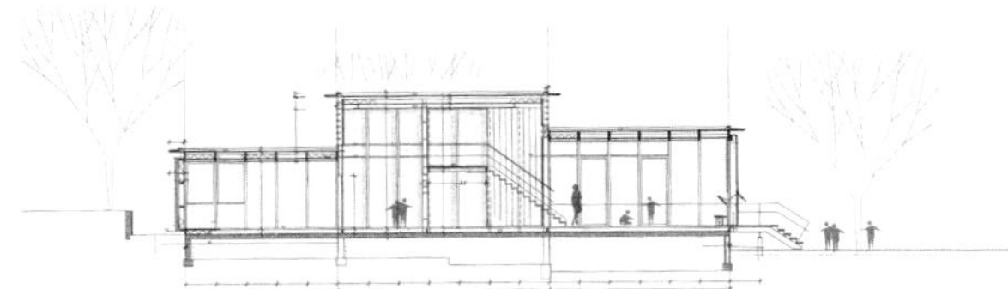

Transversal section

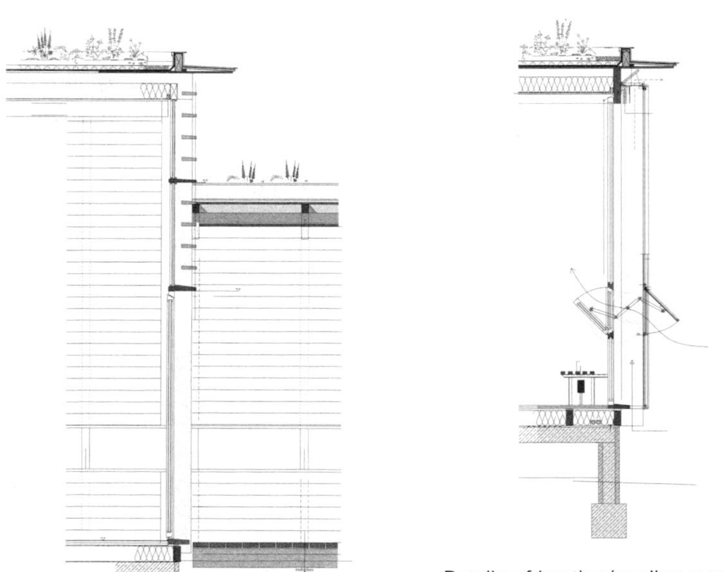

Details of heating/cooling system

Wood, in addition to being the building's structural material, is used to face sections of the exteriorand the interior finishings. The different applications create a formal arrangement that suits the design concept and the goal of the construction. The lathwork has been left in its rough form, contrasting with the fine finishes of the shutters and the panels of the cabinet furniture and other elements in the play areas.

Other green strategies include the use of local materials and the low energy consumption involved, through the use of the southern orientation on one front and technological additions.

Thus, the southern exposure provides increased heat to the air pockets in the double glass façade. In summer, when the fold down windows are open, a draft is created as part of the cooling system. The draft chamber is then closed during the winter months to reduce loss of heat from the interior of the building.

The roof includes thermal insulation materials and is also equipped with the means to hold rainwater for the orchard. Any remaining rain water is used in the lavatories. Taken all together, the series of energy conserving concepts used reinforces the idea of maintaining the new structure in an urban setting with reduced impact on the previously existing system.

Naestved Pedestrian Bridge

Andersen & Sigurdsson Architects

The pedestrian and bicycle bridge over the rails of the train station at Naestved, Denmark resulted from a desing competition held in 1997 by the municipal authorities in collaboration with the highway commission and the railway agency in Denmark. After the prospective projects were handed in, in 1998, there was ample opportunity to evaluate the winning project against through the eyes of private firms. This made it possible to cut down on expenses and study in greater detail the construction processes without significantly altering the aesthetic qualities.

The two different uses of the bridge are reflected in its form, materials, and type of frame. The superstructure inside the whole complex consists of two longitudinal elements. One of these is a large convex steel beam, a closed arc serving as the main truss. The transverse section of this element uses the geometry of a quarter of a deformed circle. It is comprised of pieces of metal welded together, some rigid (the ribs) and some curved (the outer part). The closed part of the bridge is the bicycle lane and the other part is an open steel structure suspended from the north end of the main beam at 9.8-foot (3-meter) intervals. The section is made of arcs of circles, some concave, others convex in their arrangement. This part is used for pedestrians.

Architect: **Ene Cordt Andersen, Thorhallur Sigurdsson / Andersen & Sigurdsson**
Location: **Naestved, Denmark**
Surface area: **2,260 sq. feet**
Construction date: **2000**
Photography: **Sigurdur Pall Sigurdsson**

The metal bridge rests on large reinforced concrete piers of conical shape. These, in addition to the foundation, were the elements that were actually constructed on site. The piers may be read as instilling a structural rhythm: they divide the rails and demarcate, on the western part of the bridge, the connection platforms.

The facings of the two ends of the building are different. At the station square, the building has been closed with a block used for bicycle parking, just where the majority of the trajectories cross. At the opposite end, the bridge becomes a spiral ramp taking traffic in the opposite direction-to the platform.

Access from the bridge to the station platforms is through three vertical elements that intercept the raised walkway. This vertical arrangement is serviced by a stairway and an elevator. The main element of the connection platforms is the elevator shaft, a stressed concrete structure open on the sides and with a light roof of metal and wooden studs. It curves out over the elevator and the stairway exit.

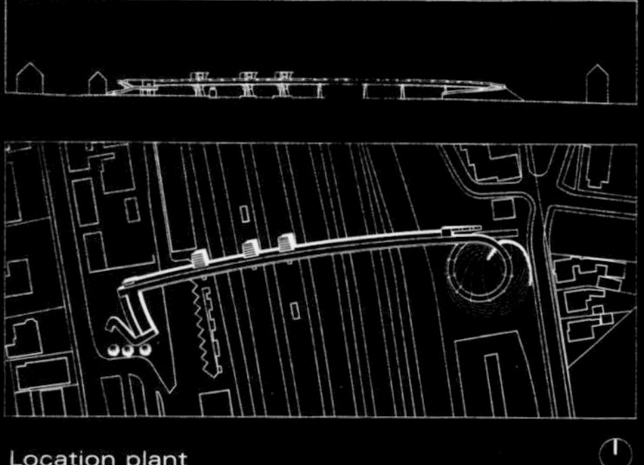

Location plant

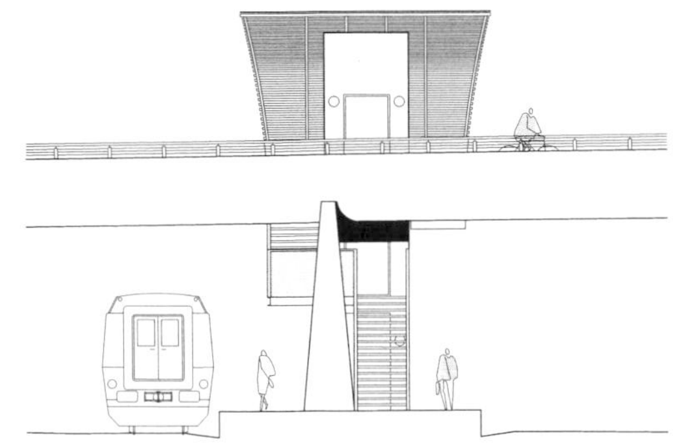

South elevation
Platform connection

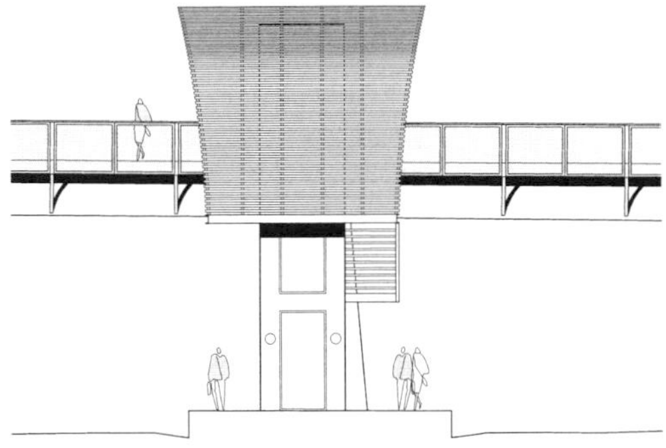

North elevation
Platform connection

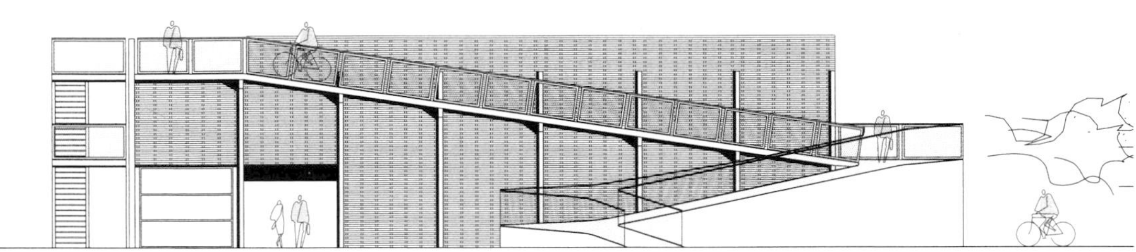

Elevation
Bicycle parking

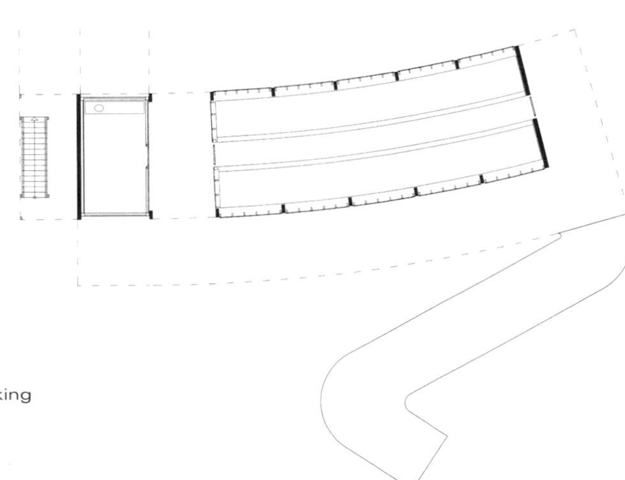

Plant
Bicycle parking

The svelte and elegant look of the bridge is achieved through a careful use of fine materials. Concrete, steel, glass and wood combine together to bring out the different textures of each material.

While the main structures in the bridge are made of concrete and steel, there is also a series of connectors made of extendable metal and larch, materials which, employed in this way, bring out the line of the arc of the circle in the urban setting. The materials used to face the parking area for bikes is wood, and the design allows natural ventilation and lighting in the space but still provides security.

The handrail on the northern part of the catwalk is made of metal frames. The elements are assembled to include sections of metal screen. The lightweight assembly on this side contrasts with the handrail on the opposite side, closed and with a continuous wooden finish. The lighting for nighttime use is held in three perpendicular pieces attached to the handrail.

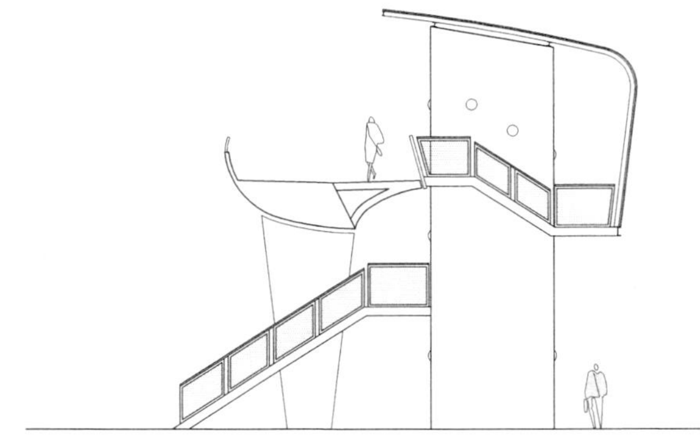

Transversal section
Connection platform

Floor plan
Connection platform

Geometry section

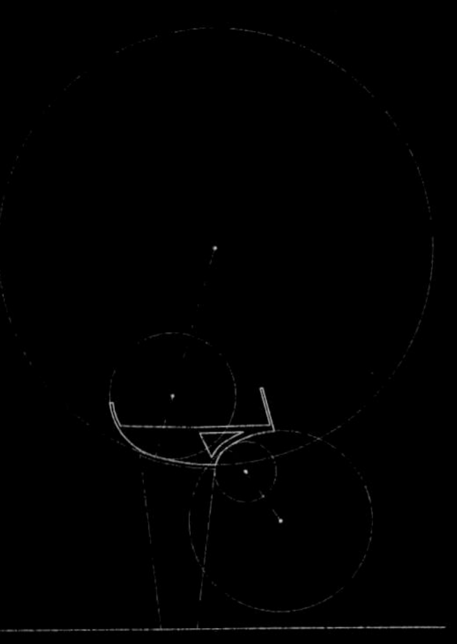

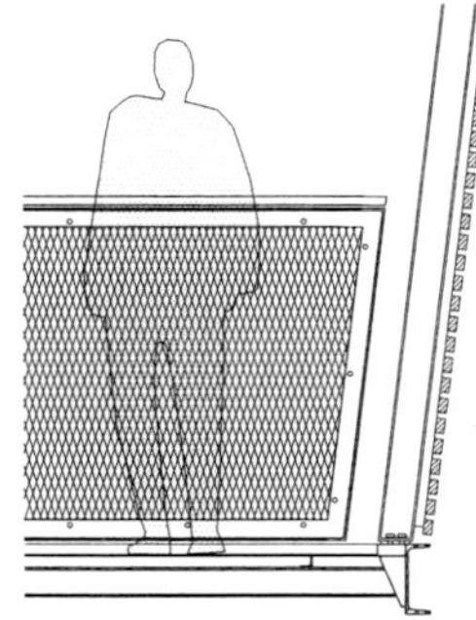

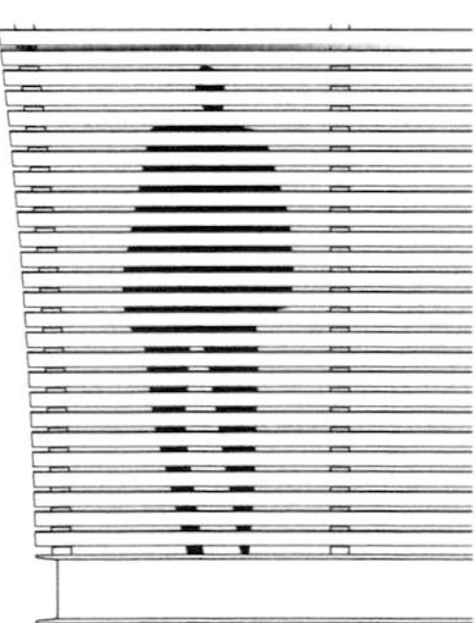

Details
Connection platform

Tahquitz Canyon Visitors Center

O'Donnell + Escalante Architects

Some centuries ago, the ancestors of the Agua Caliente Band of Cahuilla Indians settled in the Palm Springs area of California. They developed complex communities in the canyons called Palm, Murray, Andreas, Tahquitz and Chino. In the midst of an exuberant natural environment, the group thrived and grew strong on the basis of the different cultures that made up part of their economic and social foundations. Many traces of their activities, represented in art objects, construction, and engineering are conserved as a sign of their rich culture.

The project described here began as an exploration of the relationship between the Cahuilla community and the natural ecosystem in Tahquitz Canyon. Using topographical studies, field research, and interpretations of the environment, the project was developed into one of exploring the vocabulary rooted in the place and growing out of the opportunities and relations with the distant past of the Cahuillas. This point of origin then became the source of many unexpected and formal connections.

Prefabrication, or construction done in a factory, is understood by O'Donnell & Escalante Architects as a superimposing of systems. Beginning with the foundations and continuing on to the roofed terrace, a three-dimensional mesh was constructed to serve in the design and construction of repeated pieces. The variations along the Z axis, that of height, were measured and put together on site.

Architect: **Lance Christopher O'Donnell & Ana Maria Escalante**
Location: **Palm Springs, California, USA**
Surface area: **2,583 sq. feet**
Date: **2000**
Photograph: **Lance O'Donnell**

The idea of putting up a light, easily constructed building came from to cause minimal impact on the landscape. The site, located at the mouth of an old Cahuilla canyon, had remained closed to public use in order to protect the ancestral remains it held, which involved one of the oldest of the region's settlements. At the very outset, the impressive rock face of the valley walls, the main landscape feature, required a design that would respect the place, not only because of its historical tradition, but owing to the abrupt geological formation.

The view one gets from the entrance is that of a building that has been suspended over the site and embedded into it at the same time, revealing its form only partially. A short path leads to the pedestrian bridge from which it is possible to look up at the walls of the canyon or down on a small arroyo with natural flora and fauna. From there, one gains access to the Visitors' Center and the 180-degree panoramic views of the canyon and the city below, which have been curiously framed. The building is as austere as the canyon protecting it.

Floor plan

1. Entrance
2. Hall
3. Administration
4. Theater
5. Projection room
6. Terrace
7. Storage
8. Patio
9. Footbridge

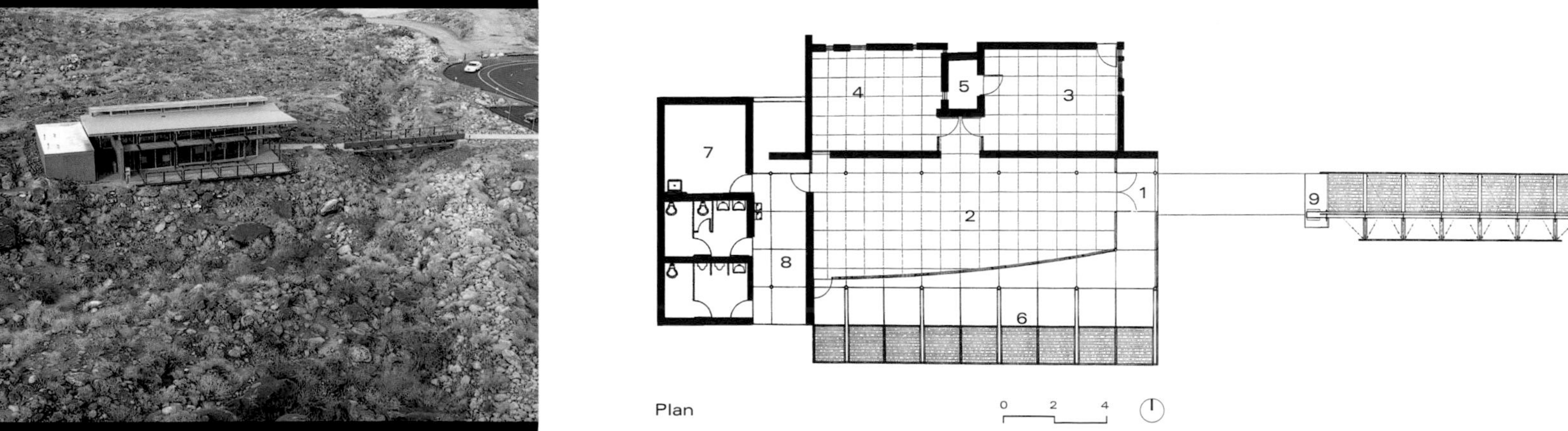

Plan

The system used to design and construct the Center enabled the architects to join the different pieces together in what may be considered a perfect fit. The glass front, for example, has exactly the same dimensions established in the blueprint, and the same is true of the wooden beams, which were prefabricated and assembled with separate metal pieces. The back of the builder's pickup truck thus had its contents reduced to concrete blocks, steel, glass, plywood, and plaster, all of which were used judicionsly and efficaciously so that there is simply nothing wasted.

Tahquitz Canyon Visitors Center acts as a slide projector that shows the Cahuilla past via a hi-tech building project that treads softly on these ancestral lands. The center features exhibition panels with maps, photographs, and maquettes. They show the extensive archeological research necessary to reveal the codes of the culture that remained hidden for 2,000 years.

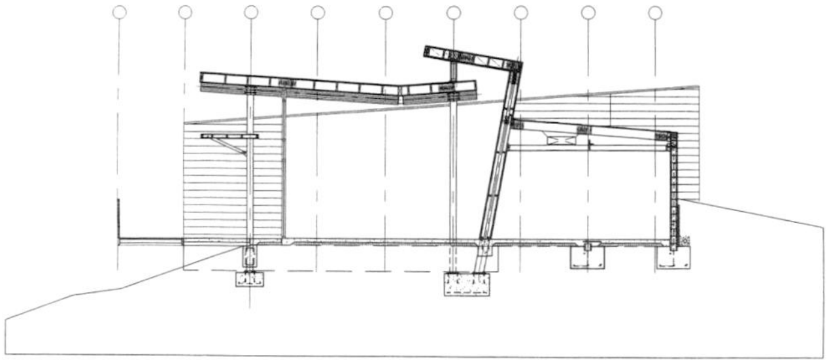

Transversal section

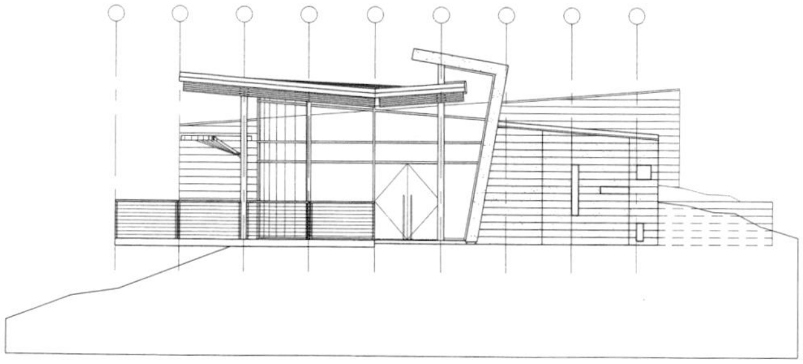

Elevation of the entrance

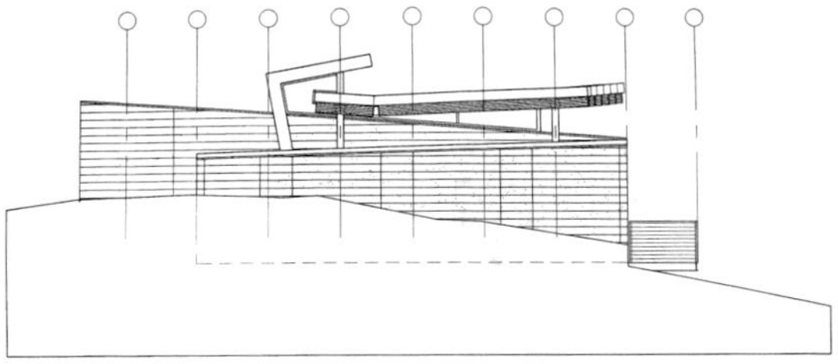

West elevation

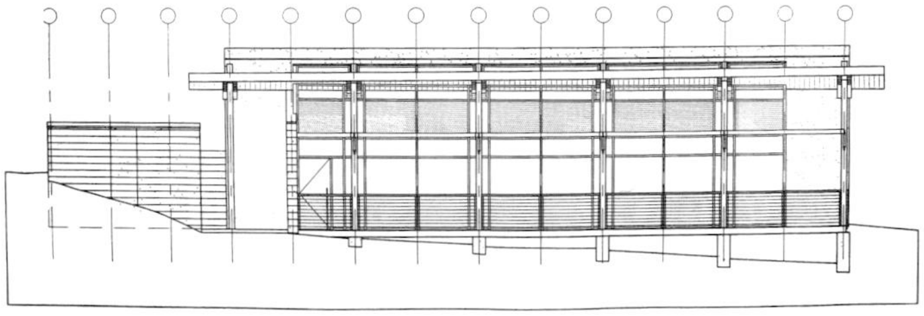

South elevation

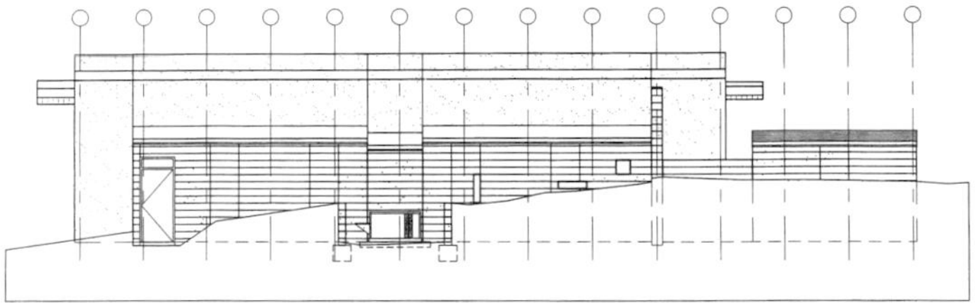

Lateral elevation

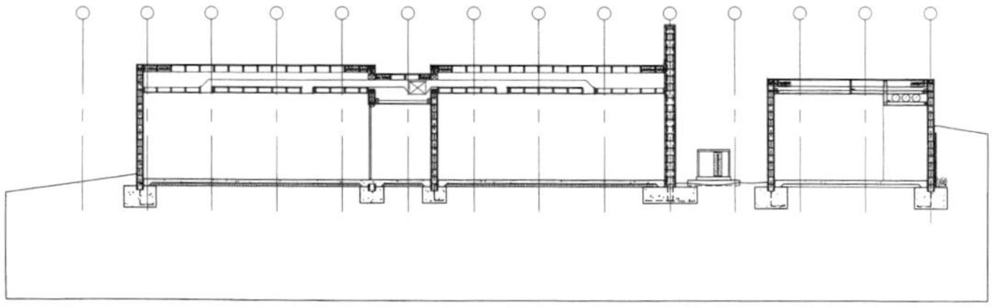

Longitudinal section

The large overhangs and the microperforated metal screens protect from the intense sun that bears down on this area. From the interior of the building, the visitor may enjoy the wide panoramic views of the surrounding landscape through the oversize windows.

G House Hans Gangoly

Designed by the architect himself and his partner, this house is based on a simple structural concept that exploits its location extremely well. The small site, with neighboring buildings on the narrowest sides, is able to count on far-off views to the east and west that clearly demarcate the house's general alignment and composition. The roof and the mezzanine floor and the two glass sides, which are very solid and, as it were, monolithic, are on a system of metal columns along the longitudinal axis. This proposition brings about a singular diaphanous interior space, free of other structural elements, which can be used with some freedom. At the same time, it reduced the time it took to build by placing structural elements and leaving the general cladding to be done minimally.

The two long façades, the main compositional and functional elements of the house, are not load-bearing walls. They are made of prefabricated frames and large-format glass panes. The frames are not inserted between the side walls but placed on top of the walls. Hence, the walls and the other planes in the house are not on the border and the defined limit between them comes out of their very juxtaposition.

Architect: **Hans Gangoly**
Location: **Graz, Austria**
Surface area: **1,230 sq. feet**
Date: **1998**
Photograph: **Paul Ott**

The large flat surfaces of glass are partially covered with upright awnings both inside and outside the house. This generates different ambiences but, principally, it guarantees an adequate lighting through the transparent fronts of the building. Moreover, in summer, energy is obtained through the interior extraction device, which is used to heat water, and to recycle cold air.

pOn the ground level, a wooden deck extends the living and dining room spaces, uniting them with the front garden on the western side. Optimal advantage is thus taken of a relatively small (1,205 sq. ft.) construction by integrating it into the surrounding landscape and creating a visual feel that is much larger. The greater part of the domestic activities take place on the eastern side of the house. One passes a monolithic staircase that joins the building with the street (the main entrance, in fact) and continues to the first floor. Finishing off this sequence is still another sculpted external staircase that takes the visitor to the roof, which is also used as a terrace. It is quite a varied and playful program, extending the interior and exterior possibilities of the house.

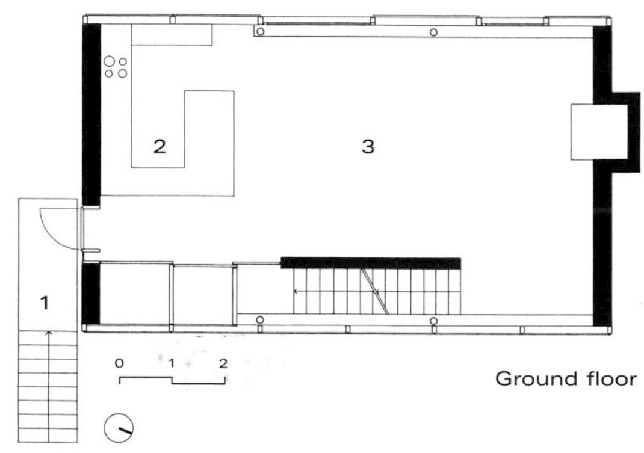

Ground floor

1. Entrance
2. Kitchen
3. Living room/dining ro

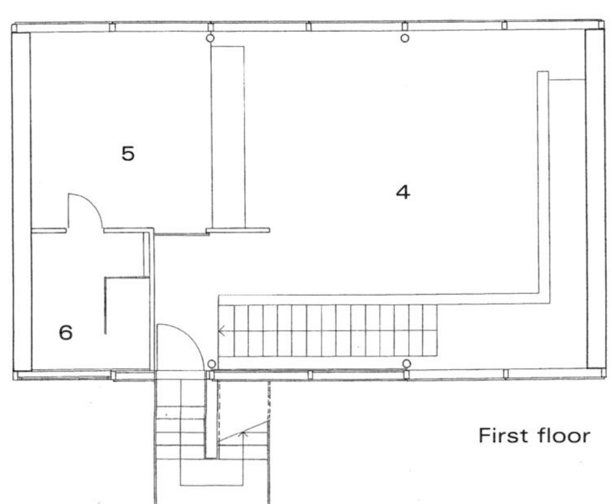

First floor

4. Studio
5. Bedroom
6. Bathroom

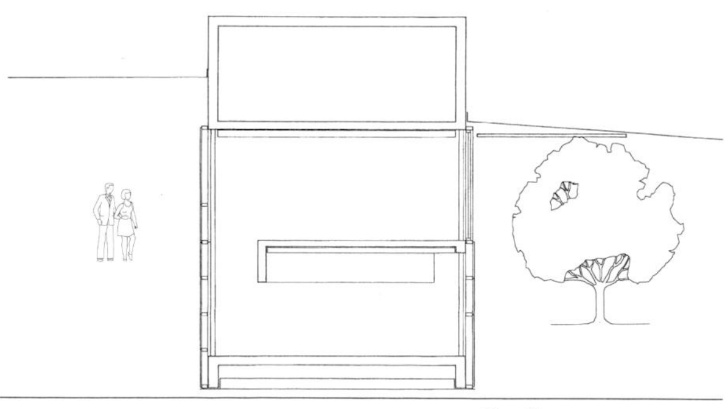

Section

The prefab framing system and that of the glass wall create a free and independent composition that interrelates with the different proportions, both fixed and movable. The vertical awnings inside and outside interface with the upper sections of the house to create a nearly transparent adornment.

The main elements on the northern and southern faces, the entranceway and the open chimney, are arranged on the building's longitudinal axis, creating a symmetry that resolves the interior distribution. The projection of the front door defines the foyer and the circulatory routes through the house. The fireplace dominates the living room and the dining room. An L-shaped space between the ground and first floors has been attractively set up and enriches the interior space.

The ornamentation and finishes have been simplified to the point of reductionism, using the construction elements themselves as final textures. The cement, metal, and glass in the general structure provide different nuances in their combinations with the wood in the floor, the staircase, and part of the furniture. The kitchen, integrated into the ground floor space, is a U-shaped area in the southern sector. To define the remaining zones, lightweight furniture emphasizing the spatial flow has been used to advantage.

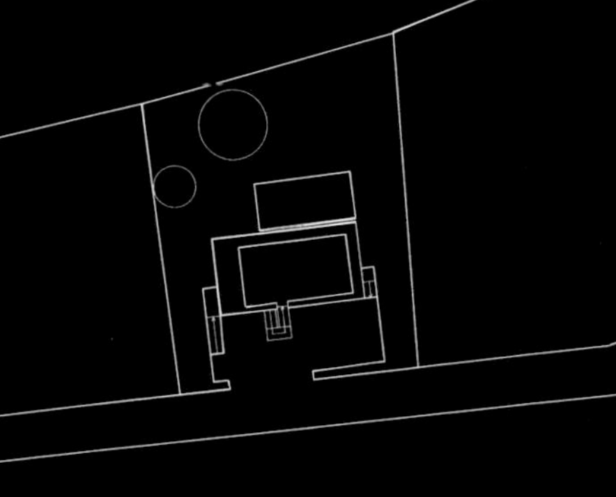

Location plan

Pen House

Querkraft Architects

Fifteen minutes by car to the west of Vienna, a compact, use-oriented house for a family of three was constructed on a more than 1,000-square-foot site. The first operation in this project was to demolish the old summer house that had been built there, dating back to post war days. A light wooden box was then set onto the prepared basement foundation. This decision, guided principally by the economics, led to the conception of a very light building that would be easy to build and, at the same time, built on off the site.

The old parking lot was kept while the sublevel was newly designed to create an area with a genuine garden. This is a solid concrete element in the shape of a U that rests on the terrain itself and creates a solid base in its upper part. A full façade of glass creates a direct interface between the interior and the exterior space.

After the stability of the basement level had been assured, the construction of the wooden part was a matter of only a single day. This was made possible by the degree of prefabrication with which the project was conceived, reducing to the minimum the parts actually constructed. The prefab wooden panels, the frames, and the window assemblies required no further adaptation once they left the factory and were transported to the site.

Architect: **Jacob Dunk, Gird Erhartt, Peter Sapp, Michael Zinner / Querkraft Architects**
Location: **Mauerbach, Austria**
Site area: **1,033 sq. feet**
Date: **2001**
Photography: **Hertha Harnaus**

The compact geometry of this house is the result of skillfully exploiting the property in order to gain free space around the building. The orientation of the site, with the garden on the southwest side, is ideal for setting up the interior. At the same time, the location in a narrow valley makes it necessary to move the main living area upstairs, from which the finest panoramic views are obtained.

The size of the balconies that run along the front depend on the functional needs of each floor. On the ground floor, where the bedrooms and the studio are located, the balconies work simply as a light extension of the space. But on the upper story, where the living rooms, dining room, and kitchen are, the balcony is large enough to fit a dining table.

The lateral railings and ornamental work of the balconies are supplied by a low-cost netting, of the type used in sports facilities.

Avoiding rigid dividing walls, the project distributes the space on plan as open areas promoting great flexibility. The location of the staircase contributes to the various definitions of public or private rooms. On the ground floor, the rooms may be separated or integrated at will; on the upper level, vertical movement marks the boundaries between the living room and the dining room.

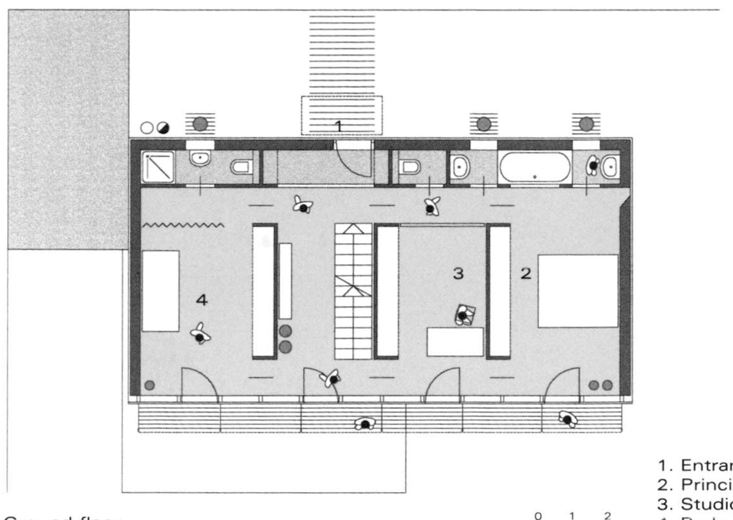

Ground floor

1. Entrance
2. Principal bedroom
3. Studio
4. Bedroom

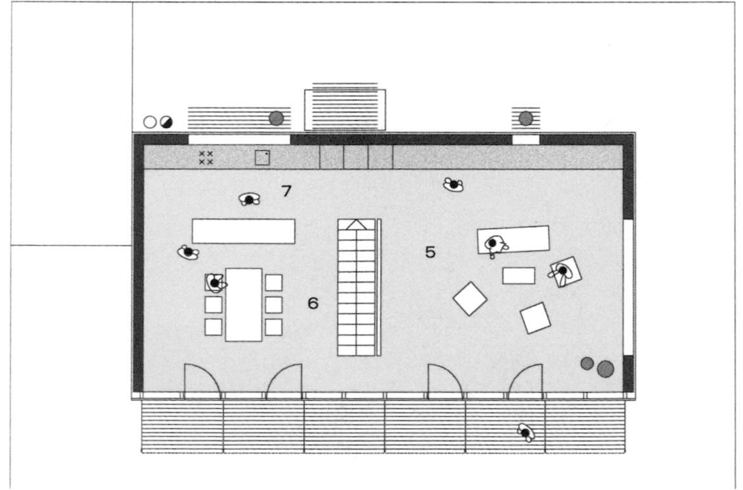

First floor

5. Living room
6. Dining room
7. Kitchen

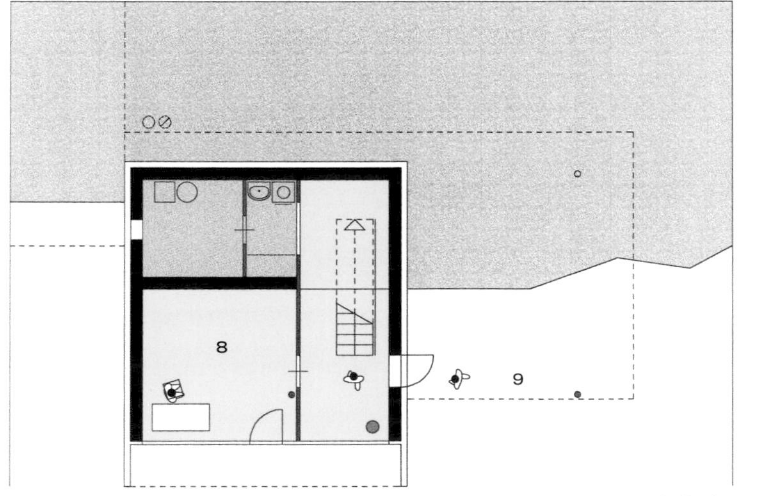

Besement

8. Bedroom/Studio
9. Garden

A narrow service area, just under three feet wide, contains the service rooms in the back of the house. This arrangement takes maximum advantage of the inside dimensions.

Location plan

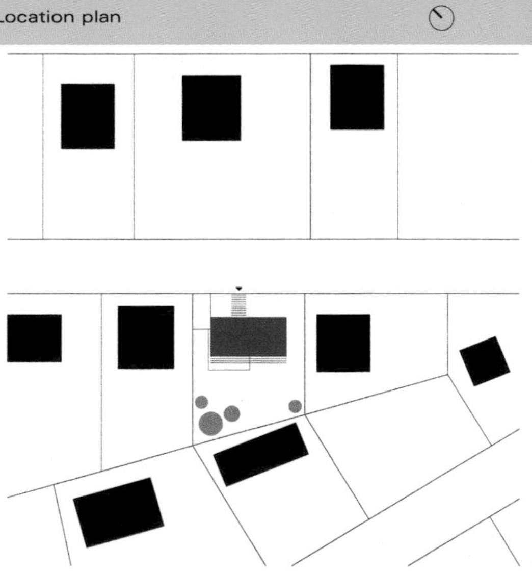

Six small iron balconies dot the main front, hanging like delicate complements to the hard-line architecture employed. Each of these contains a pot of flowers that livens up the effect, like a prolongation of the front garden. The building is entered by a lightweight wooden door that is at street level-the level of the ground floor-and that negotiates the natural rise in the terrain.

The eastern façade has a play of windows on a plan that would otherwise leave closed some of the wooden panels. A perpendicular opening divided into different small sections to filter in the morning light demarcates the main room of the ground floor, and a horizontal slit on the upper story looks out onto the view of Vienna from the living room.

Although practically all the elements of this residence are prefabricated or industrialized products, the building carefully responds in a particular way to the specific conditions of both the urban and the natural setting.

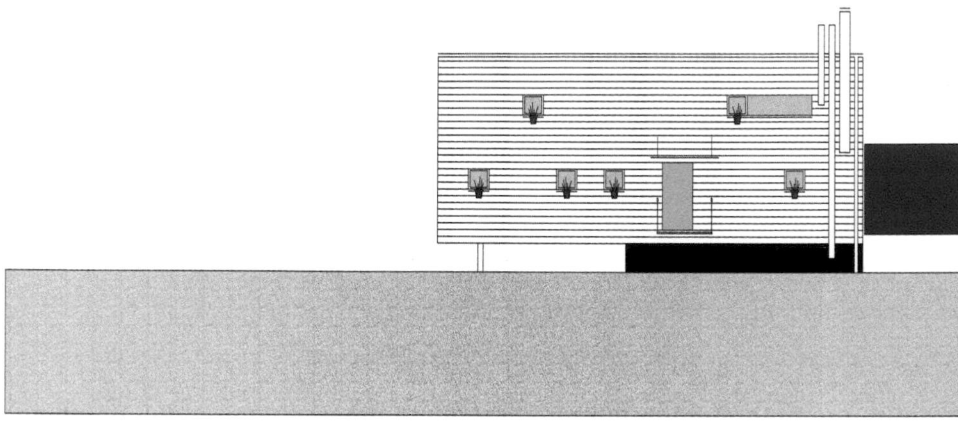

North elevation

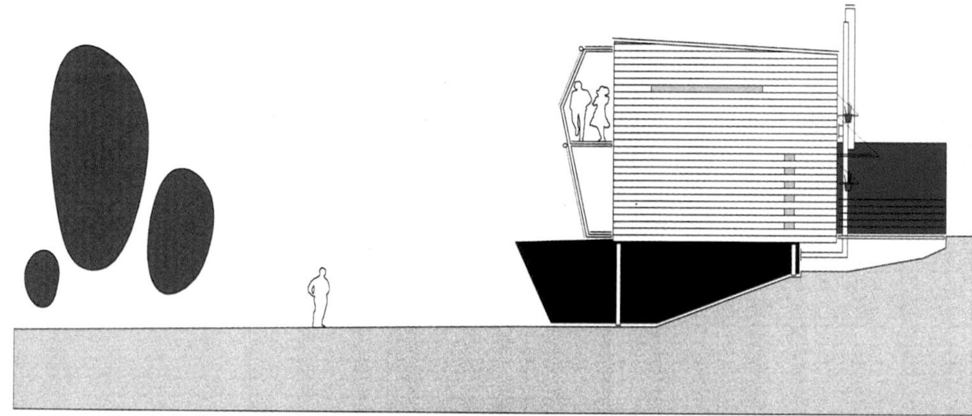

East elevation

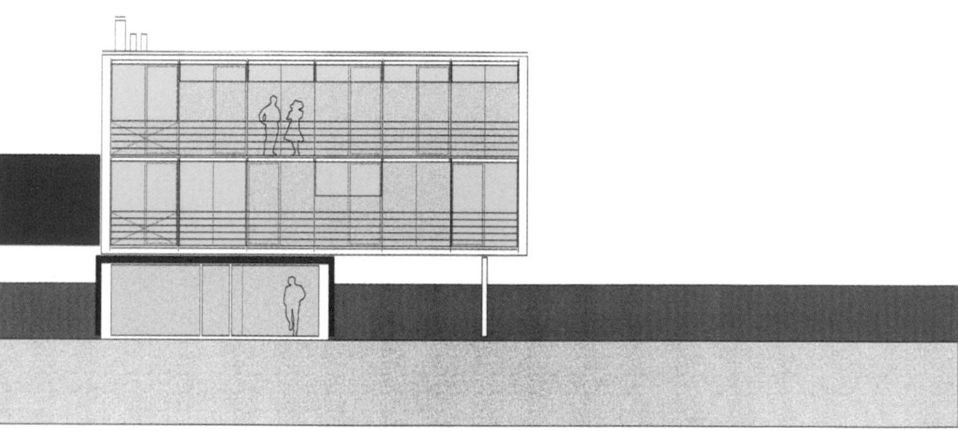

South elevation

5
Betreten der Baustelle verboten!
querkraft

Tea House Marterermoosmann

The architecture of this small building is a response to different parameters affecting its construction. On the one hand, there were the requirements imposed by the client, a tea taster from Hamburg who resides in Austria and who was seeking a quiet place, simple and pure, and befitting his profession, hence the name of the house. On the other hand, however, the difficult conditions involving the site's accessibility (the steep slope on the northern face, however splendid its views) were what finally defined the construction. The aim was to achieve something minimalist that solved the problem of openness, lightness, and honesty.

From the word go, the builder collaborated with the architects in the design process and the search for the most appropriate solution both in terms of cost and design. The technique used would be a major element in the house's formal result. The decision to put together a prefab building was made mainly because of the difficult trip up to the terrain for heavy vehicles: the construction would have to be done with only two heavy equipment operators and in a very short space of time.

The first step was a reinforced concrete base to define the grade in relation to the slope and simultaneously create a solid mass on which the rest of the structure would rest. To avoid excavating below the point where the ground was frozen, thermal cement was used as a foundation. Onto this a structure of three stories was raised, using metal rods eight feet apart.

Architect: **Georg Marterer, Thomas Moosmann / Marterermoosmann**
Location: **Neustift am Walde, Austria**
Surface area: **1,615 sq. feet**
Construction date: **1997**
Photography: **Manfred Seidl + Marterermoosmann**

The light metal grid was planned according to these dimensions, as were the assembly system of aluminum framing pieces, including the sliding doors and windows and the facings of the building. On each floor, the structure was reinforced by revealed metal supports that anchor the building diagonally and are part of both the finished exterior and interior. The cladding alternates fixed glass panels, sliding windows, and Siberian birch wood panels in three layers with a PVC core to provide greater insulation. Inside, the panels are finished in plaster and the windows are air-filled double panes.

The intensity of the natural lighting inside the house is controlled by using a variety of systems. There are the blinds that affect both interior and exterior transparency. There are wood panels that work on rails as well as other planes that are suspended horizontally on graduated cables, trellis-like, joining the front wall in the kitchen area.

The play of contrasts is a constant feature inside the house: the cold materials of the structure over and against the front wall of warm wood. There is an open interplay between the closed and open façade, between the intimate and the extroverted. The few interior walls there are do not fully extend to the ceiling, giving the space a greater fluidity and flexibility.

Detail of the assembly of metallic pieces

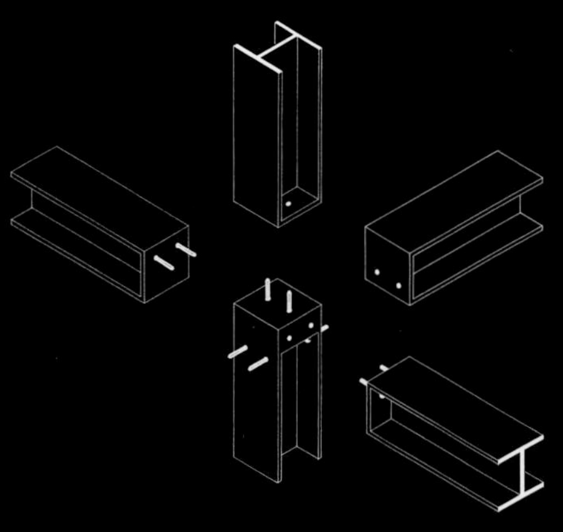

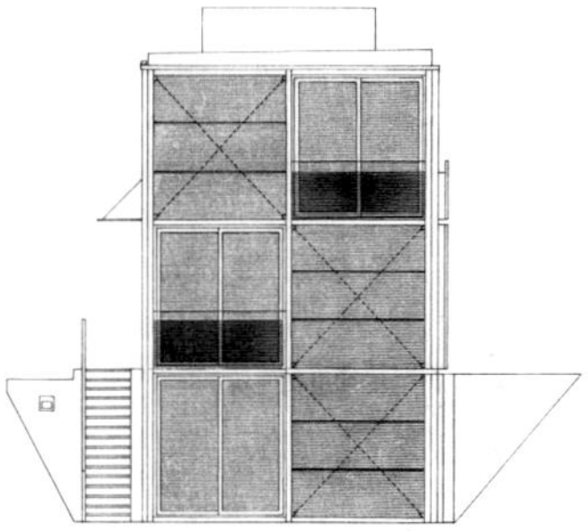

Front elevation

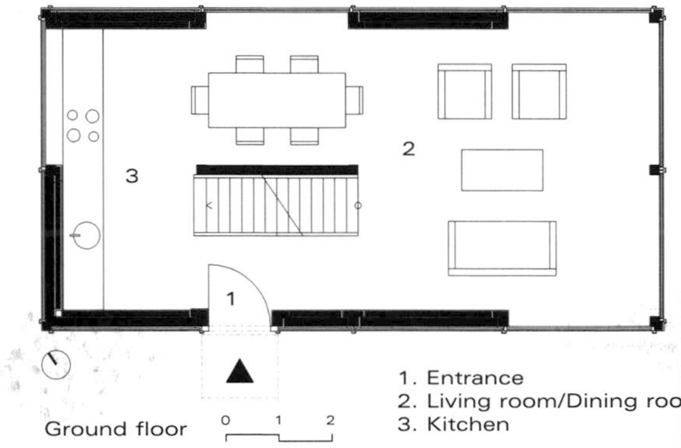

Ground floor

1. Entrance
2. Living room/Dining roo
3. Kitchen

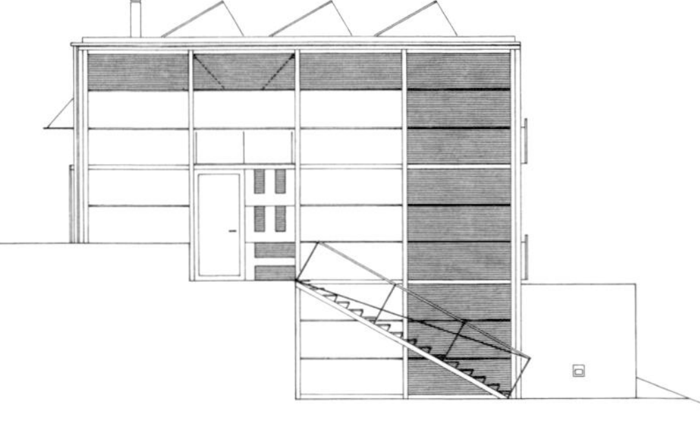

Lateral elevation

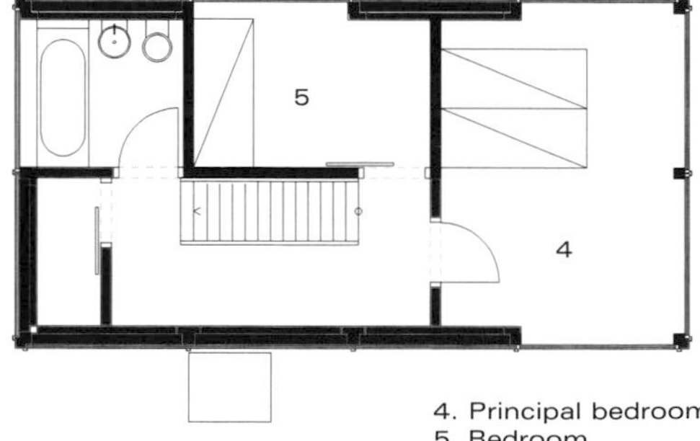

First floor

4. Principal bedroom
5. Bedroom

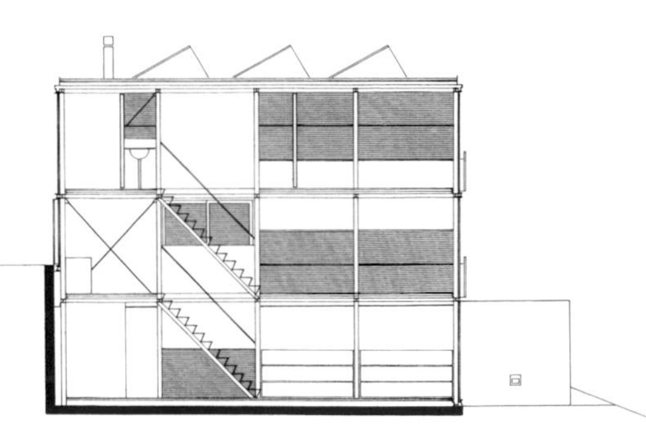

Longitudinal section

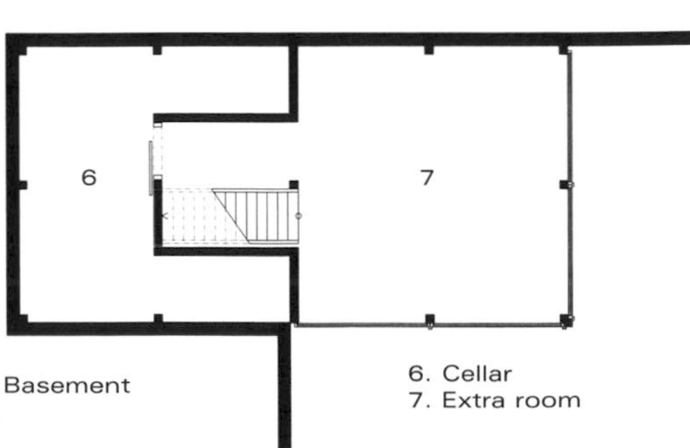

Basement

6. Cellar
7. Extra room

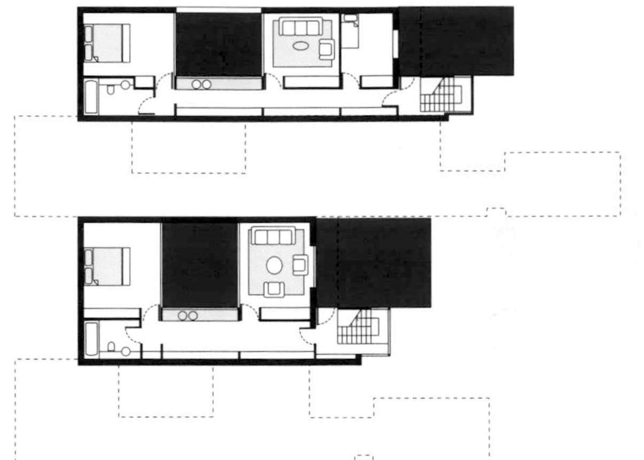

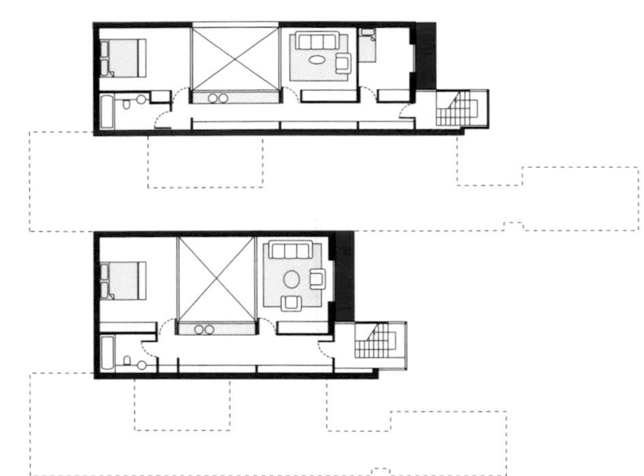

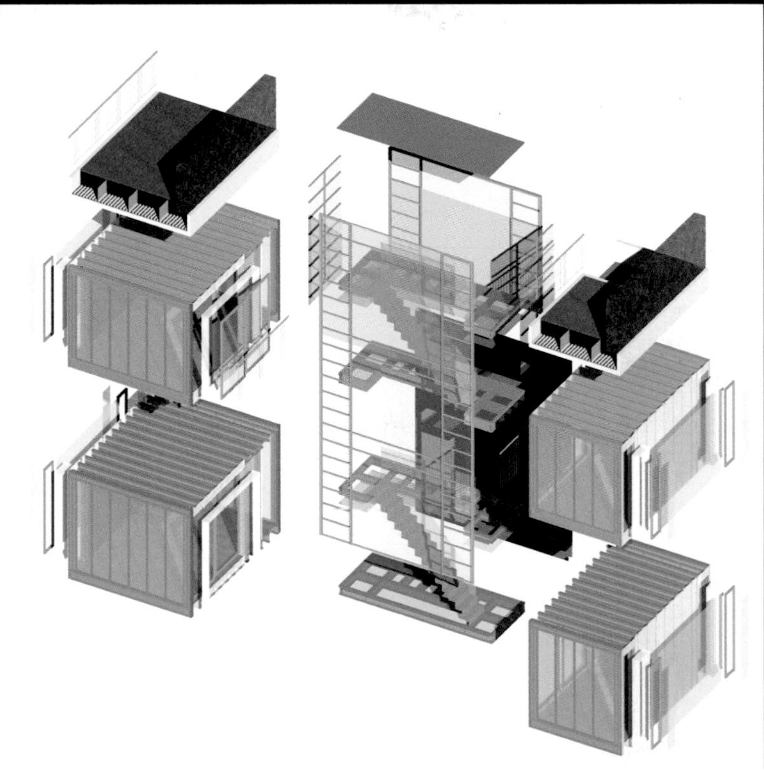

An urban scale model conceived down to the last small interior detail of the dwelling, with customizing left to the residents themselves.

Architect: **Arthur Collin** Location: **London, United Kingdom**
Surface area: **1.000.000 sq. feet** Construction date: **2000**

Wooden Modules

The project described here would create a new and extensive development area in East London based on the potential of the setting. The idea: generate a greater feeling of responsibility for public spaces and create a more sophisticated hierarchy between public spaces and private spaces.

Perhaps all public spaces are multifunctional, but this project has avoided ambiguous zones, promoting greater density and assuring the use and maintenance of public areas. Hence, we see a conflation of high-rise and low-rise building typologies creating a dense urban weave of great variety and character.

The houses are assembled from prefab wooden frames comprised of columns and I-beams. The structures thus becomes lighter, with the same resistance afforded by solid wood and at the same time possessed of easier traffic flow and access to space and installations. The master walls incorporate all of the house's basic elements and provide completely fitted units.

Inside, the composition can be varied by its inhabitants, adding or eliminating modules to adjust to private needs. The units can be shaped either as small bath-and-kitchen space or else as a traditional multiple-roomed house.

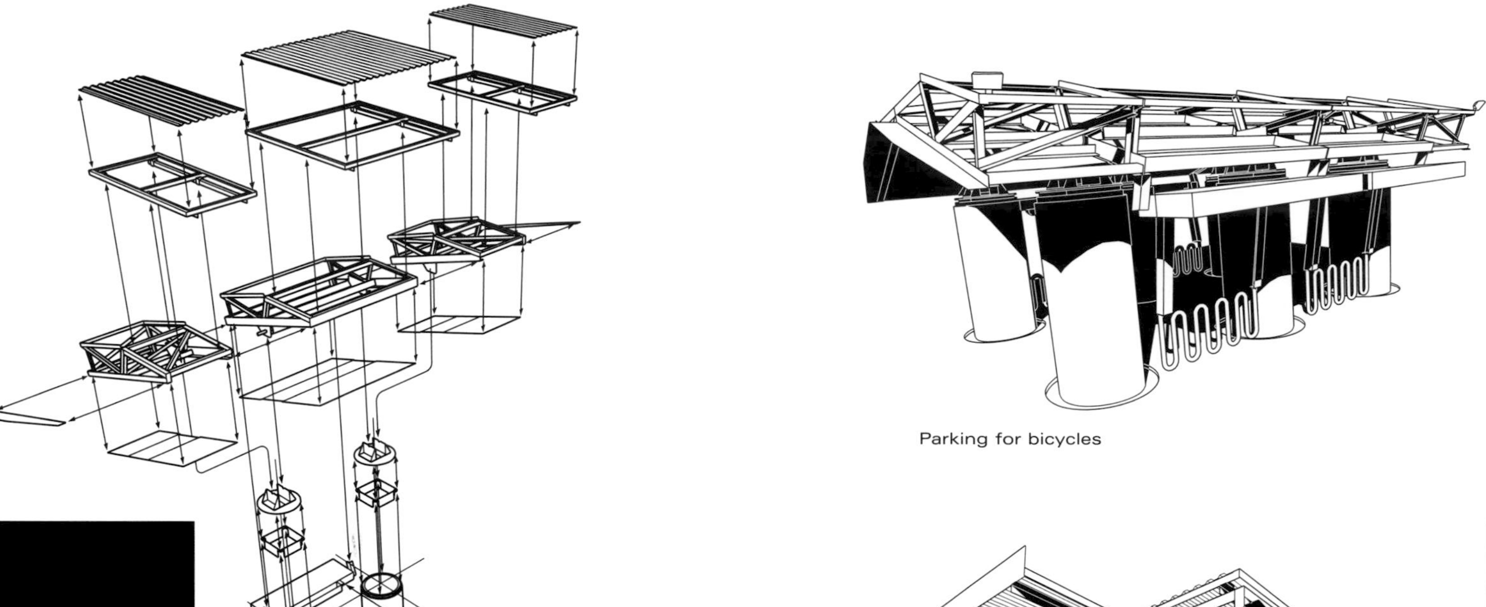

Parking for bicycles

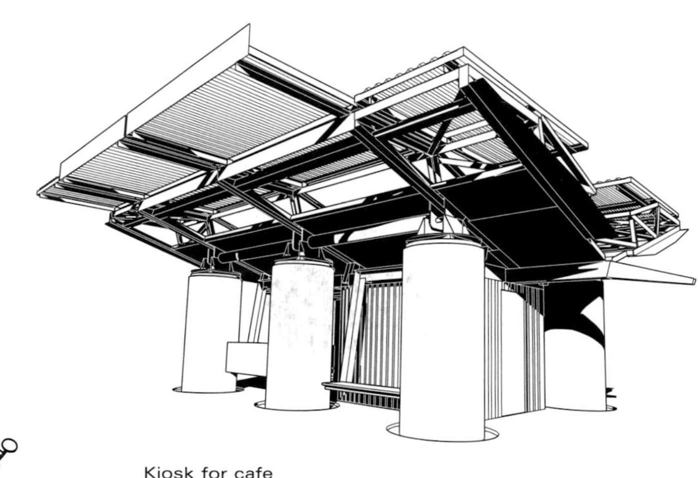

Kiosk for cafe

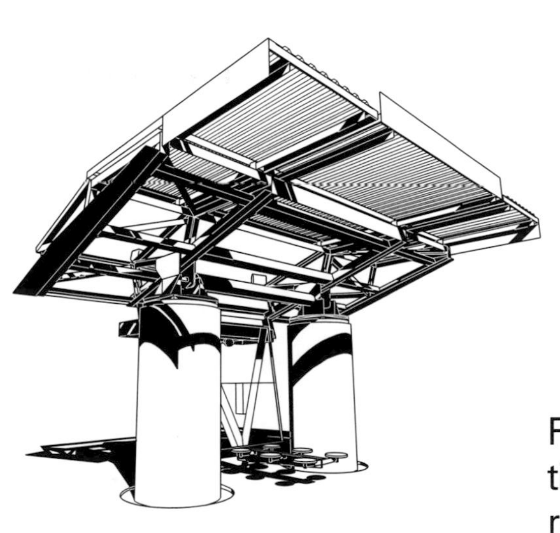

Bus stop

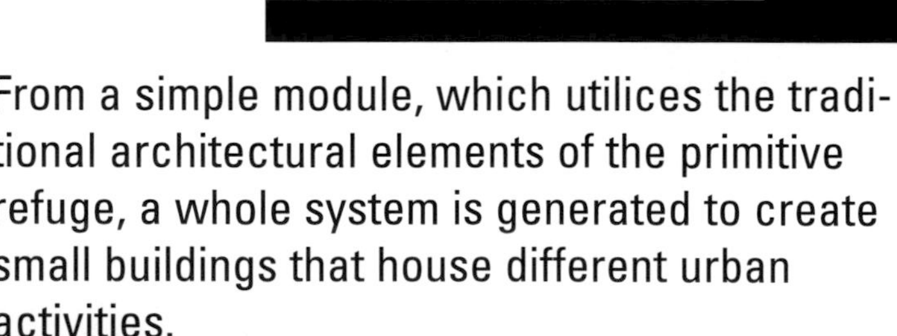

From a simple module, which utilices the traditional architectural elements of the primitive refuge, a whole system is generated to create small buildings that house different urban activities.

Architect: **Jones, Partners: Architecture** Location: **Stanford University, Palo Alto, California, U.S.A.** Surface area: **107.6 to 1,076 sq. feet** Construction date: **1996**

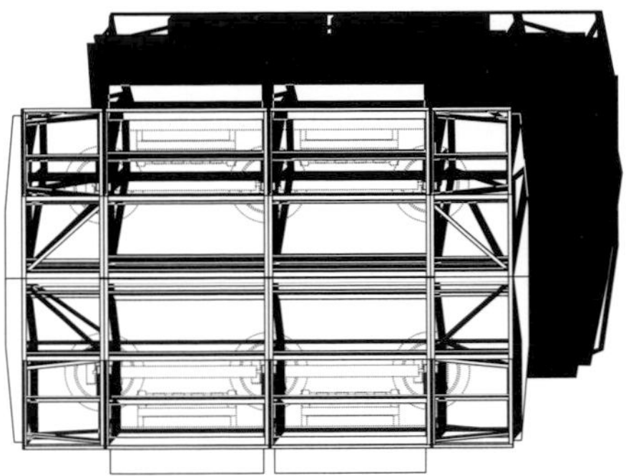

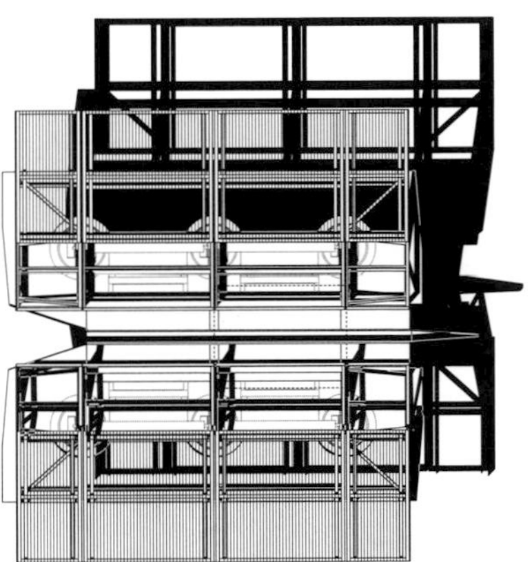

Modular Shelters

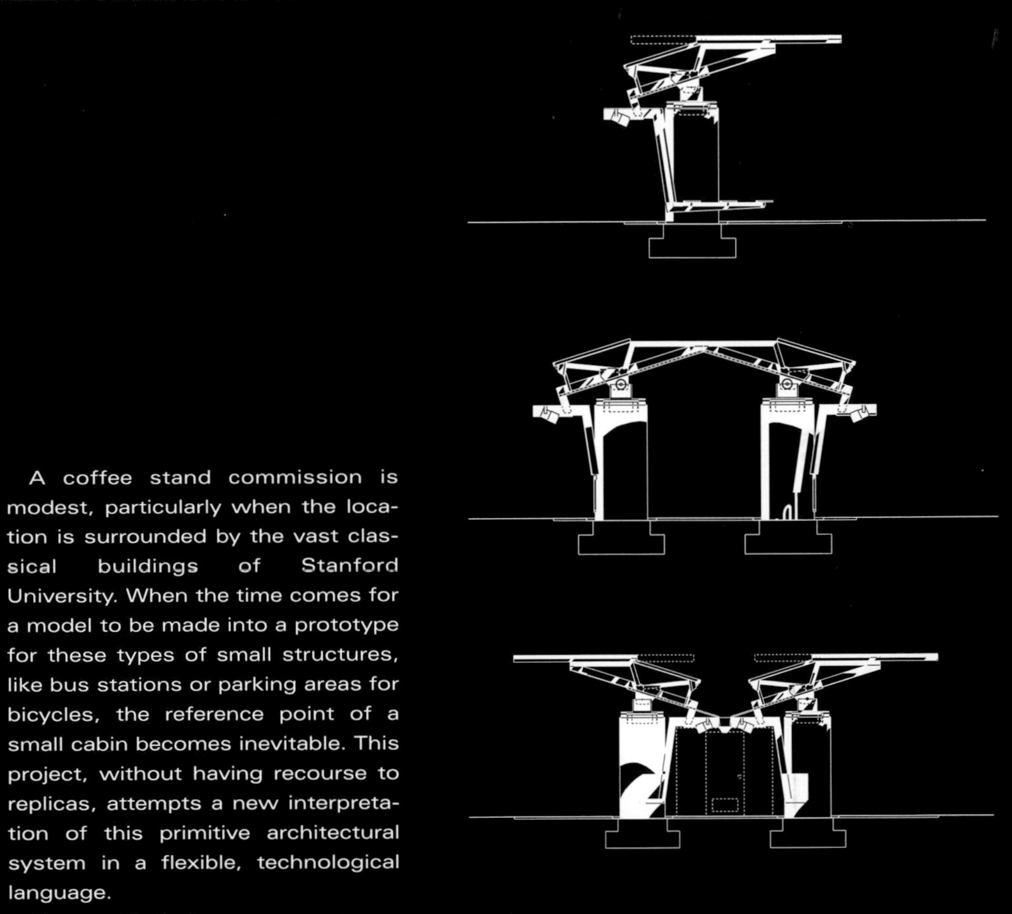

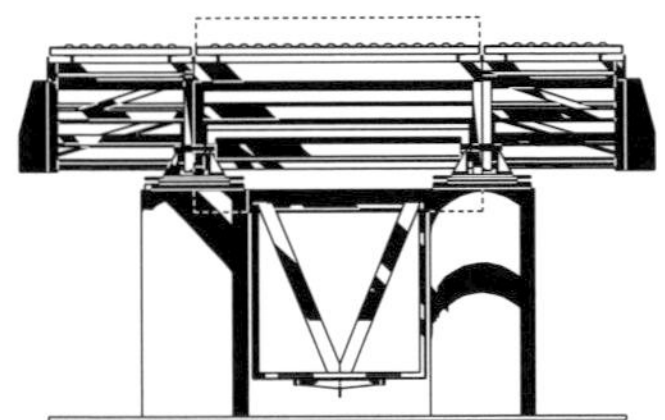

Bus stop. Elevation

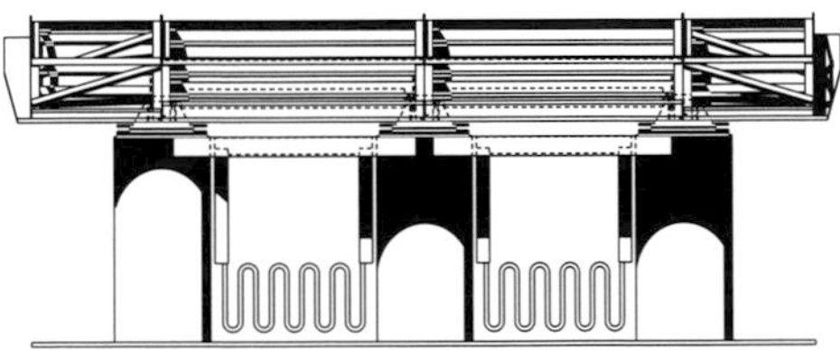

Parking for bicycles. Elevation

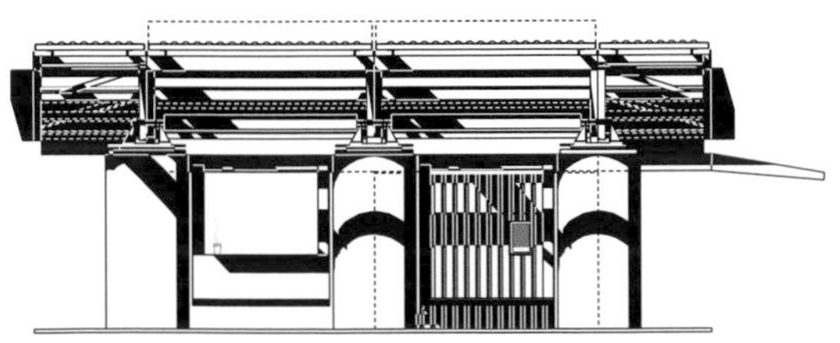

Kiosk for cafe. Elevation

A coffee stand commission is modest, particularly when the location is surrounded by the vast classical buildings of Stanford University. When the time comes for a model to be made into a prototype for these types of small structures, like bus stations or parking areas for bicycles, the reference point of a small cabin becomes inevitable. This project, without having recourse to replicas, attempts a new interpretation of this primitive architectural system in a flexible, technological language.

The small shelters are an expansive distillation of the traditional elements that go into the making of a primitive cabin. The exercise is one that lays special emphasis on creating spatial boundaries that demarcate a refuge. The different can be desingns drawn from minutte variations in the basic module-the coffee stand-from which the rest of the system derives.

Subway Stations

Despang Architekten

Metro D-Süd subway line construction work was the place where the project described here began. It would become the main transportation artery between Hannover's city center and the Expo 2000 trade fair area. The challenge was to generate a varied series of twelve stations and to have each one set up a relationship with the specific urban scene and at the same time to have the series contained within one single formal and functional unit.

The design starts by breaking the station up into a series of objects strung along the rails. A set of full and empty spaces is created. The surrounding city environment is moved inside to make up part of the imagery of each station. This modular combination of floating steel platform as six waiting-station blocks at each stop gave rise to a mass-production system and standardization of the architectural elements.

The strategy results in minimal construction costs and maintenance, on the one hand; on the other, it serves to back up the orientation and recognition of each section of the subway line. The spaces built for waiting passengers slot themselves effectively into the city, expressing their specific personality via variations in finishing materials. For travelers who use the line, this underscoring of the specific personality of each station is a kind of choreography, dramatized and condensed in the direction of the last stop, at Expo 2000.

Architect: **Despang Architekten**
Location: **Hannover, Germany**
Surface area: **161 sq. feet (each module)**
Construction date: **2000**
Photography: **Despang Architekten**

Adaptable Modular Dismountable Light Mobile

The security-both of the elements themselves and of the passengers who use the subway line-is part of a general approach. As it is a question of public service, careful attention was given to the inclusion of all of the service parameters in order to optimize use and durability. All of these are integrated into the waiting area modules, thus guaranteeing their protection. The design offers protection from the weather conditions and from vehicular traffic. At the same time, there are safety factors involved in having open-air stations with excellent visibility.

The Clausewitzstrasse vicinity is dominated by brick buildings that date back to a number of different periods. The station modules were thus designed to include pressed brick facings with half-inch mortar joints. A transparent anti-graffiti coating has been used to protect the surface from vandalism. The Kronsberg station, faced with large pebbles embedded into cement panels, recalls to mind the city center in Wülferode, which has been preserved.

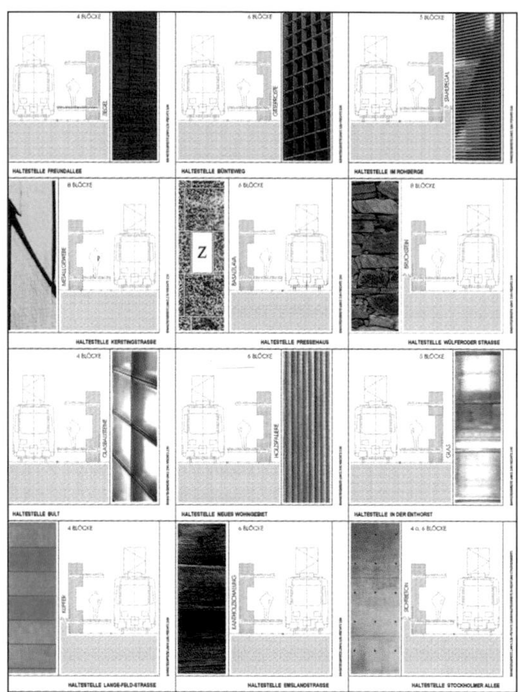

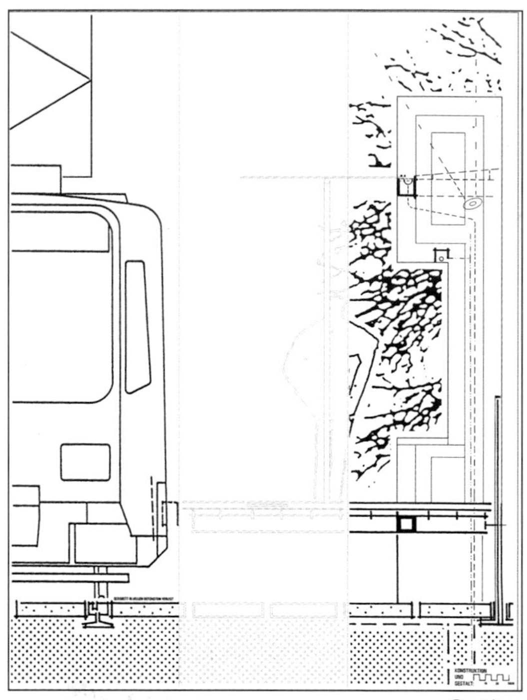

Section

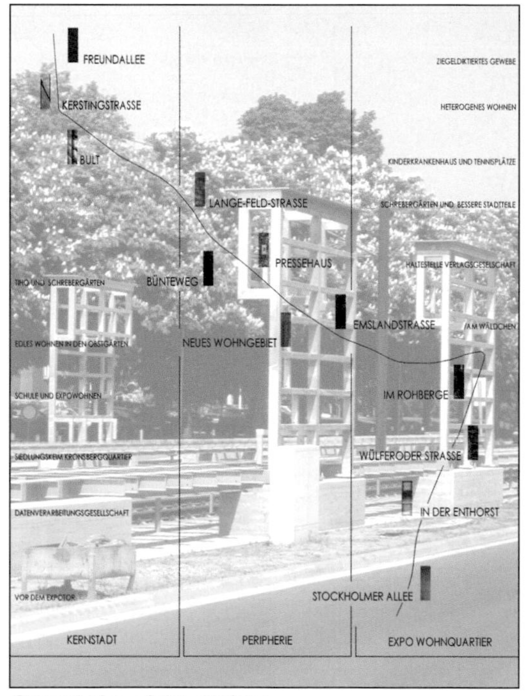

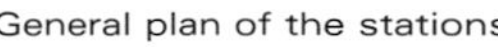

General plan of the stations

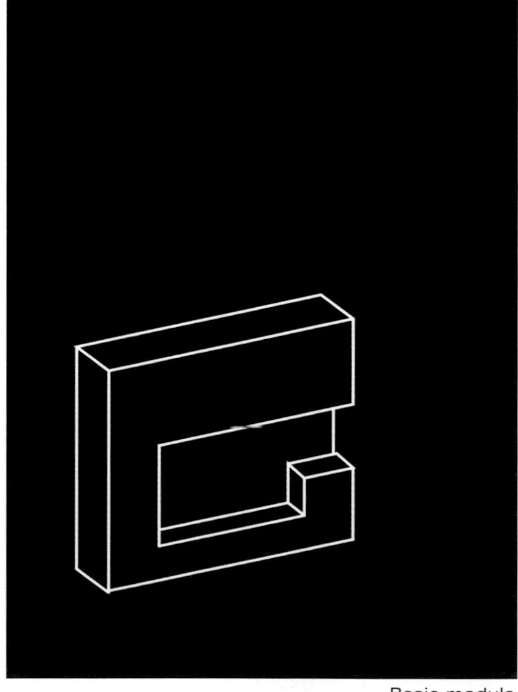

Basic module

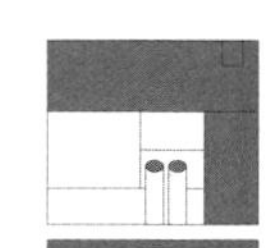

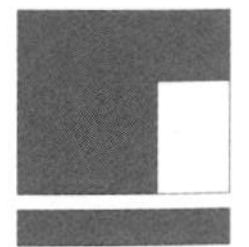

Different compositions

Kronsberg

The modules consist of a steel framework that contains the different materials. On one of the sides, the frame is stepped back to insert a large bench into the structure. The frame is stabilized by being anchored to the steel-and-concrete structure of the station platform.

Each wooden bench is flanked by two perpendicular 3/4-inch panels of security glass and another in the upper part that serves as a roof, again as a provision for weather conditions. The roof is held in place by a tubular frame anchored to the platform structure.

The minimum thickness of each element allows maximum visibility of the rails and makes it easy it see the incoming train. Information on itineraries as well as advertising are carefully installed in each module and can be removed for consultation as necessary.

The material employed in the furniture of each top is glued to the basic steel structure of the frame. The choices of finish, which were chosen to blend with local features, serve in each case as light upholstery and as a sign of identification for residents.

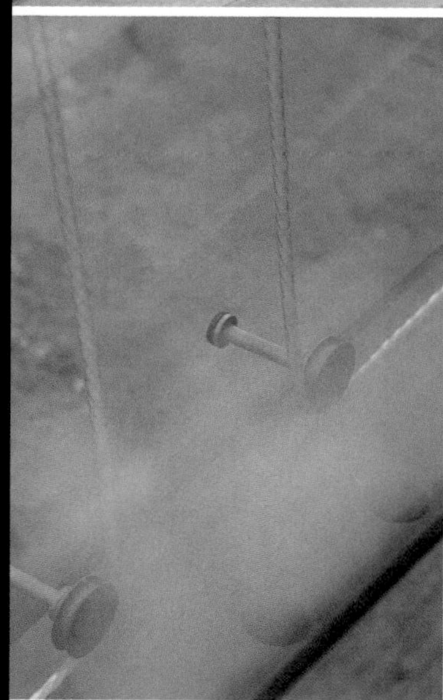

Two-Family House

Oskar Leo Kaufmann and Bmst Johannes Kaufmann

This house consists of two independent units with a capacity for one family in each, one on the ground floor and the other on the upper floor. Each of these sections is based on an identical program that includes a kitchen, a bath, a living room, a dining room and two or three bedrooms.

The construction used a 16.4 x 16.4 foot wooden module system in a stacked, side-by-side arrangement. A finished wall was added later, having been previously prepared to contain the whole building. What the system achieved was a space-saving construction that could be raised in a minimal amount of time. It is slotted into the location to form part of a single piece that becomes an adaptable dwelling able to suit the particular needs of each client.

Each house has the capability of varying in shape, size, roof type, window type, and exterior line. The 16.4 x 16.4 foot (5 x 5 meter) squares can be grouped as desired. This determines the building's configuration while also making it possible to choose from among ten different façade types for the final appearance. The distribution and proportion of the interior spaces (kitchen, bath, living room, bedrooms) can be selected according to the way each owner decides to decorate and tailer the place to personal needs.

Architect: **Oskar Leo Kaufmann and Bmst Johannes Kaufmann**
Location: **Andelsbuch, Austria**
Surface area: **6,997 sq. feet**
Date: **1997**
Photography: **Ignacio Martinez**

The façade is predominantly transparent. It responds to the features of the terrain, with its striking visual impact on every side of the house. The oversize apertures, in the form of sliding doors and sliding windows, translate into an excellent integration of the surrounding landscape. The façades include pine wood slats on the horizontal used as a device that sets off the large openings.

The door, on one side, is covered in methacrylate panels. This is the main perpendicular element that has been incorporated. It serves as a formal and functional link between the two living units. In spite of being a very singular type of architecture in this region, it has integrated itself into the surroundings on the basis of the use of the same materials that have been used there for many years.

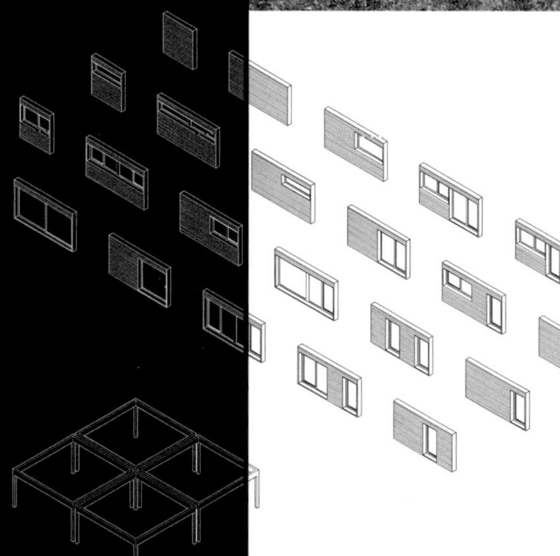

Diagram structure and façade panels

The interior is designed as a single sweeping line. The space is ample, largely because of the uncovered framing system. Rooms like the kitchen and the bath are built in the center of the dwelling and act as elements around which the space is distributed. The kitchen is also prefabricated and has been installed on the spot, permitting size and shape adjustments. The electrical system is channeled into a single central conduit running through the space between the modules' load-bearing columns.

Wood is again the predominant material in the finishings and the extra interior ornamentation. The false ceiling is made of plywood panels; the floor is medium-size parquet. The improvement in the integration of all the elements together is notable, as is the gain in space owing to the revealed skeleton. Some of the furnishings, those in the dining room, for example, or the table in the living room or the kitchen, play up this same idea.

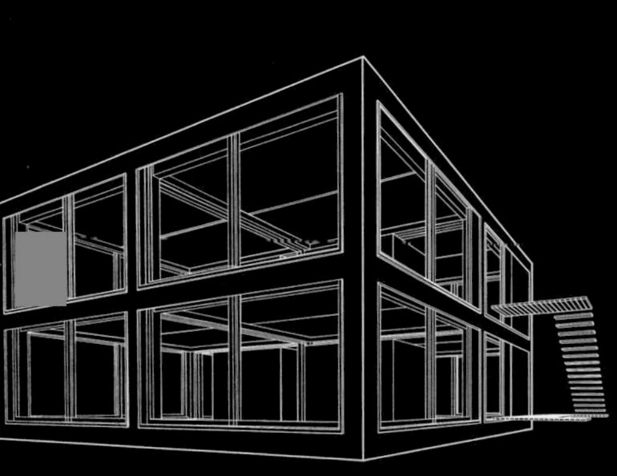

Perspective

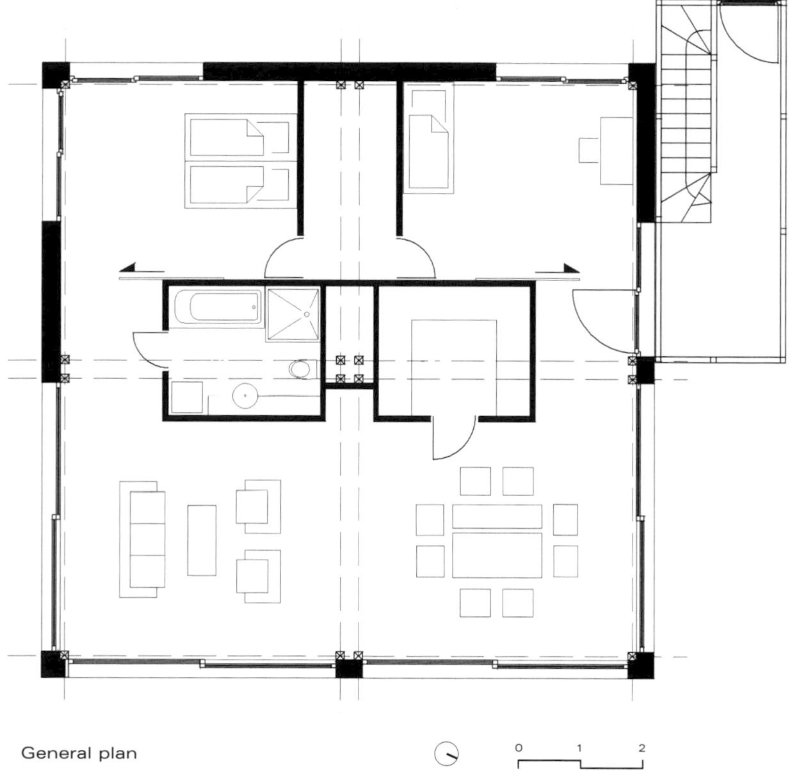

General plan

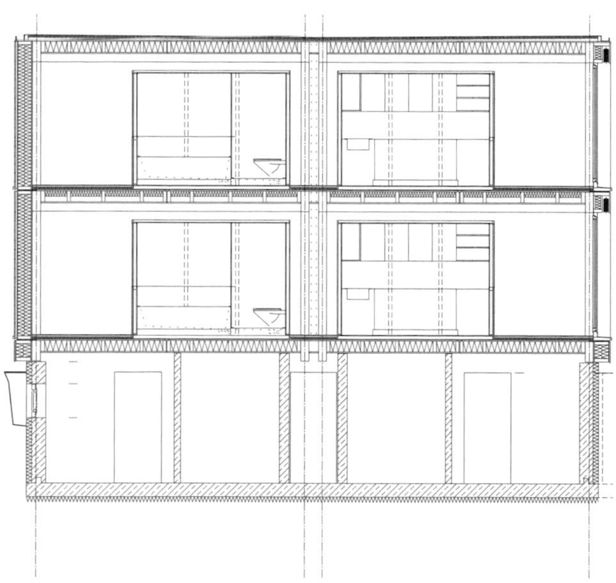
Section

The modular construction permits different distribution arrangements inside the house. There are dynamic additions outside as well, well-suited to the oversize apertures on both levels. The wooden deck was added on one side of the house by taking advantage of the sliding doors, again adding space.

House in Munich

Christof Wallner

Installed behind an already existing house, this building in the garden solves the problematic situation of the site by way of a simple plan and a light, entirely prefabricated construction. The young couple who commissioned the job wanted to install a quality addition without having to complicate either the form of the piece or its price. Their garden site is behind a thirty-year-old house, however, and this complicated its access as well as the availability of natural light to the new annex.

Pre-cut construction meant making the work zone accessible and also cut down on work costs and time. No more than four months were required in the refurbishment. The new structure employs a wooden main frame, fiberboard panels, and metal-frame windows all, as mentioned above, prefabricated and transported to the site as required.

Special attention was given to the ecological aspect of the building in order to control the energy consumption in the house. The panel construction, both interior and exterior, minimized the skin section to take full advantage of the site space and yet create sufficient thermal insulation. In addition, local construction materials were used, again to avoid long-distance hauling complications.

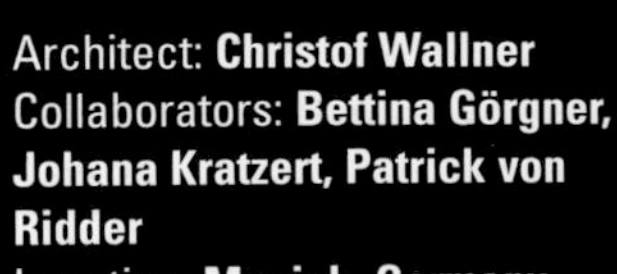

Architect: **Christof Wallner**
Collaborators: **Bettina Görgner, Johana Kratzert, Patrick von Ridder**
Location: **Munich, Germany**
Surface area: **1,345 sq. feet**
Construction Date: **2000**
Photography: **Michael Heinrich**

The project was approached with two main premises: on the one hand, impact on the already existing house should, for budgetary reasons, be minimized; on the other hand, the garden space occupied should also, for social reasons, be kept to a minimum. The answer is the cubical two-story piece which, despite its somewhat contemporary monolithic look, reflects and interprets the neighborhood's original forms and proportions.

The direct connection with the garden space was achieved by conceiving a very large glass surface to define the ground floor façade. From here, a wooden deck goes back to create a transitional zone between the greenery and "indoors." Interface with the previous architecture comes through the upper level and the catwalk joining the two pieces. This underscores t[illegible]y of the design. The house is thus a kind of double which not only shares common living areas but also foments the link with the green environment-and the neighborhood's traditional focus.

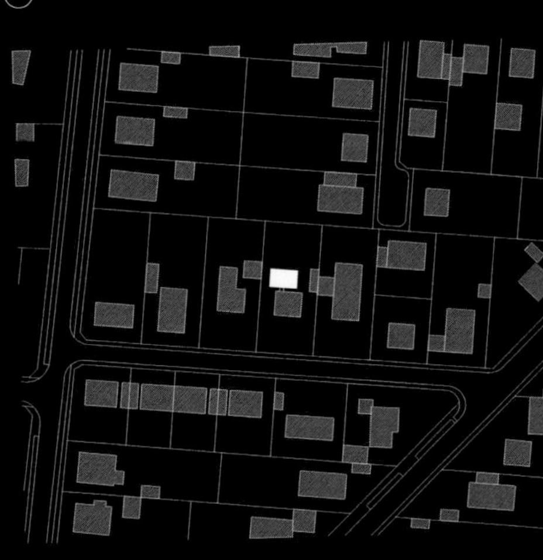

Location plan

The building's facings are of fiber cement panels varnished in a gray-blue tone. The precise geometry of the addition is highlighted by the grid created by the joints in the panels of this precise cladding. The building's apertures use varnished frames that are a dark matte gray. They appear as simple extensions of the fiber cement panels and thus reinforce the idea of a continuously flowing skin. Similarly, wood-slat shutters have been installed to protect the bedrooms in the upper story. The living room, dining room, and kitchen, on the ground floor, can also be closed off from the outside by use of the sliding doors.

Indoors, the idea of flexible austerity grows stronger through the use of light tones and lightweight materials. The white walls, sliding doors, and simple light lines of the shelving are reinforced by the natural stone floors.

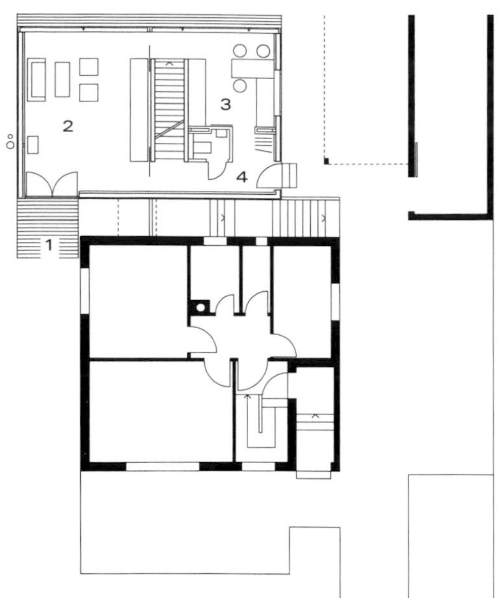

Ground floor

1. Entrance
2. Living room
3. Dining room/Kitchen
4. Bathroom

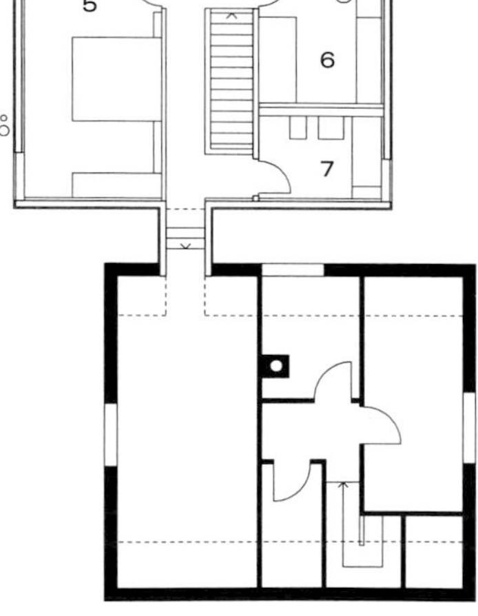

First floor

5. Principal bedroom
6. Bedroom
7. Bathroom
8. Basement

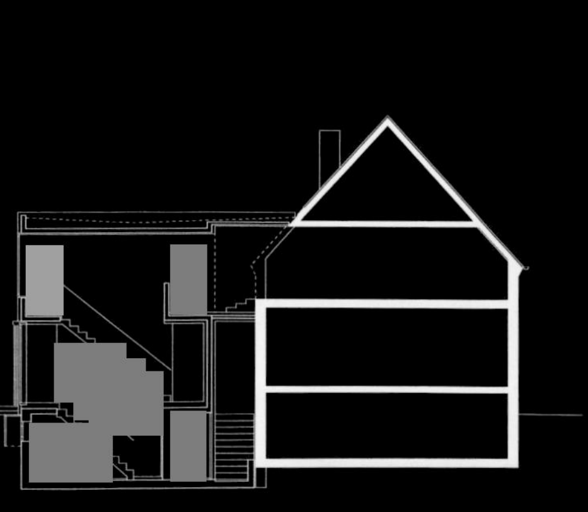

Section

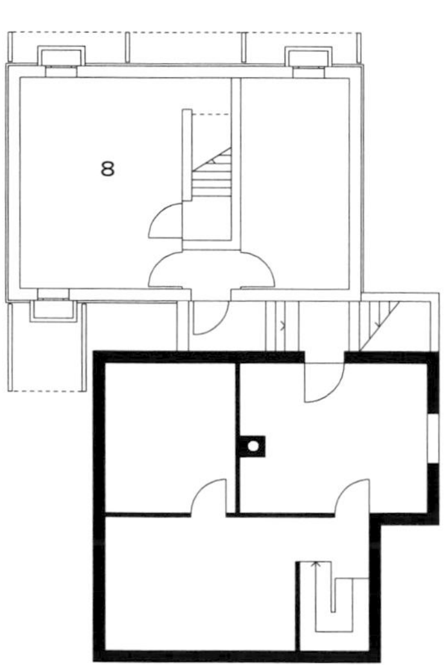

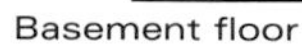

Basement floor

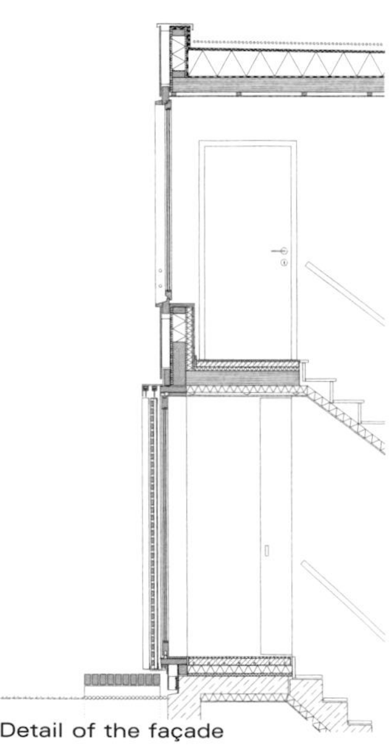

Detail of the façade

The contrast between the new construction and the reformed element highlights the way both living styles and building styles have changed.

Stockner House

Wolfgang Feyferlik

In the beginning, this project for a second residence for a family with four children consisted for two separate buildings. The first of these was was intented to be the main house; the second, smaller and complementary, was intended to be an alternative area for the changes of a big family, as a small guest house, or as a working cabin that was independent from the main house. At the time of writing, only this last part has actually been built.

The program's simplicity, the rural landscape surrounding it, and the guise of the secondary structure brought the architect to realize a measured work, something light that would minimally impact on the surroundings, even during the time of building. Prefabricated wooden frames structure and compose the house.

This skeleton is finished inside with birch wood panels, faced on the outside in rough-cut pine. The roof, however, contrasts by way of the zinc panels used on its surface. And the south façade is divided in two, with folding glass windows on one side and black-painted radiators to collect solar energy to heat the house on the other side.

Architect: **Wolfgang Feyferlik**
Location: **Tainach, Austria**
Surface: **646 sq. feet**
Construction date: **1992**
Photography: **Paul Ott**

Adaptable Modular Dismountable Light Mobile

Now, the blend of materials, wood on three faces in contrast with a metal roof and a south front in glass, creates its own salient language, halfway between the rustic and the contemporary technological.

The site itself, a gentle slope that provides splendid panoramic views out over the valley, is still visible beneath the building, which is raised slightly on wooden piers. This gesture emphasizes the lightness of the house and of course also leaves free a crawl space to store firewood.

The three sides that are practically closed, and the southern orientation, provide all the advantages of using the sun's rays to optimize heating and, through storage, to cool the interior. The windows can be fully opened. This yields well-lighted interior space and natural ventilation. The door is on the eastern side, one of the short sides, and it is a small arrangement made of rustic wood.

Although the materials and the formal discourse used in the project delineate an architecture just as typical as the rest of the buildings in the area, the house is also perfectly integrated in the surrounding environment.

The solution for distributing the floor plan, which holds to the same general concept, satisfies the needs of one or two people, depending on how the interior elements are used.

Two practically identical spaces are lighted and ventilated naturally and give onto a neutral area. This corridor serves to connect one or two rooms through an arrangement of sliding separators. At the same time, the space is wide enough to be used as a workroom, a kitchen, or a small living room.

The bath-room is a small cabinet lined in waterproofed plyboard, and its every inch is put to use. The wall-mounted toilet makes it possible to conceal the tank behind the wall and also liberates the space above, making cleaning easier and opening up the space.

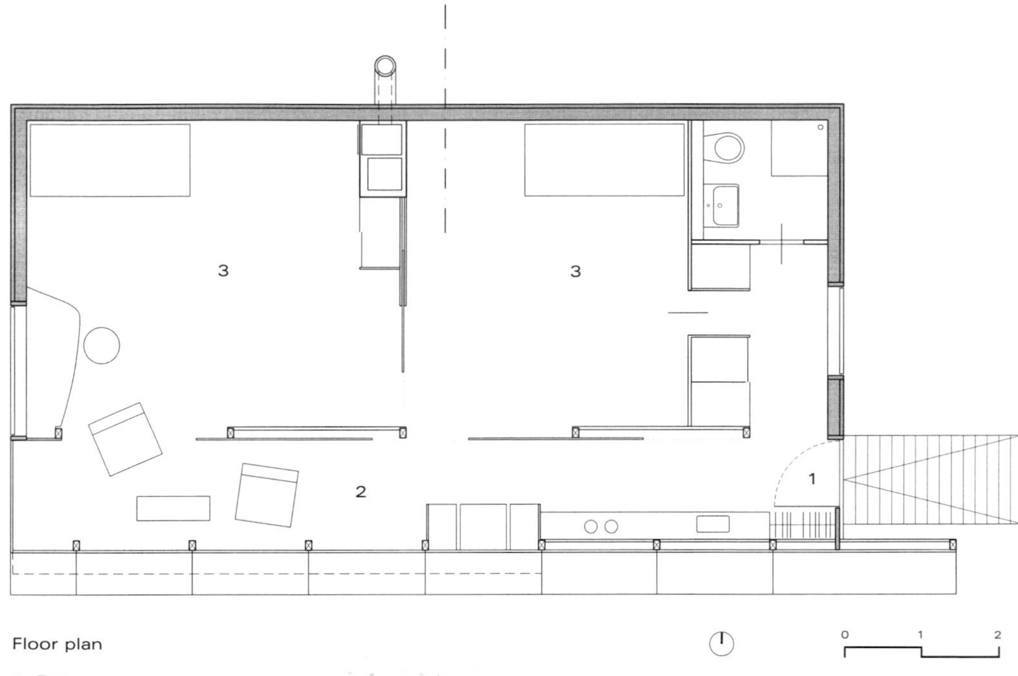

Floor plan

1. Entrance
2. Corridor/Living room
3. Bedrooms

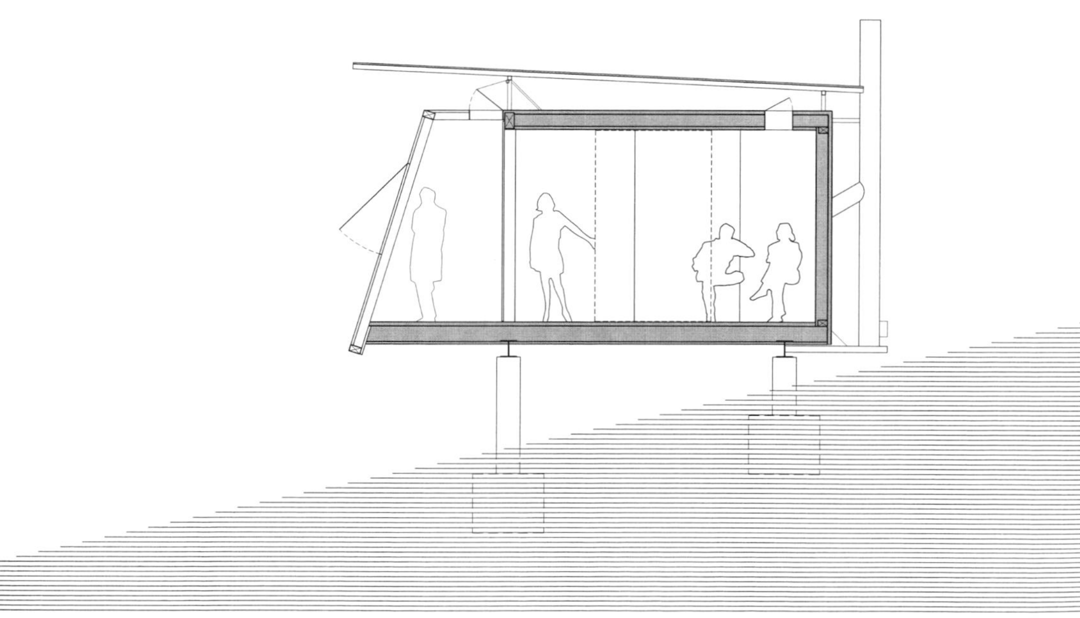

The predominance of wood inside the building accentuates the lightness of the cabin. It also makes the place warm and welcoming. Unification of materials generates a greater spatial flexibility, and a sense of depth.

Hornegg Complex

Hans Gangoly

The complex comprising the three Hornegg houses in Preding, in southern Austria, acquired its present appearance following different interventions over the course of its history. Built in 1875, the work of the engineer Daniel von Lapp (the author of the Lieboch-Wiesbahn train line), the complex denoted the first phase of the building. Its formal aspect is strongly marked by reminiscences of the type of architecture so popular at the time. Hence, we find a main block in rectangular form, with a series of minor constructions added to it over time. The refurbishment plays up the industrial character of the building in the suite of windows, the revealed brick façade, and the austere and homogeneous nature of the design.

In the spring of 1997, the attic of the main building was practically destroyed by a fire. The incident was used as an opportunity to create new living areas inside while the rehabilitation and restoration work proceeded. Additionally, the clients were interested in generating a new architecture and in a contemporary and individual language that would complement the extant structure while using to advantage the particularly privileged position of the whole piece. The panoramic views that may be enjoyed from the upper parts of the building required a project that would use 360 degree perspectives to look out over the town, the Hornegg palace, and the picturesque surrounding landscape. To enjoy these views but at the same time to provide protection from the weather conditions in the region became the starting point in the project's development.

The operation thus consisted in adding a new section which at some points even has a double height. It is, at the same time, of a great lightness, speaking not only aesthetically but structurally. The chosen construction system was one of prefabrication that made it possible to change only minimally the construction onto which the new parts would be added and to avoid competition with the original character.

Architect: **Hans Gangoly**
Location: **Preding, Austria**
Surface area: **6,888 sq. feet**
Date: **1998**
Photography: **Paul Ott**

In the design, the existing structure provided the base onto which the new edification would be installed as a lightweight independent object. This clearly allows the geometry of the refurbishing work to form an integral part of the old without fully respecting its lines and its walls, a detail that points up the independence of both old and new.

The system of metallic beams and columns installed behind the façade draws a line in the sand in relation to the freedom of this structural element. Contrasting with the brick wall of the previous architecture, with its minimum of apertures, the façade essentially makes a statement in regard to its flowing, all-covering glass cladding and the way the whole outline of the building is to be read. It is a gesture that makes it possible to come closer to the surrounding landscape that lies beyond the gaze from the living room. At the same time, it reflects--from the woods, from the town in which the building is situated--the structural system and the spatial configuration of Hornegg Complex.

The elements needed for the façades came from aluminum sheets that would wrap the glass surfaces and drop down over the upper part. The folding materials generate an array of possibilities that vary the external perception and allow different definitions of the nature of the spaces contained inside. The whole accurately devises a double height that may be used to great advantage in each living space as an alternative room, as leisure space, or work space.... The communication with the terraces on the roof from different parts of the house reinforces this freedom of movement.

The upper bays are set back from the main block. They give the appearance of glass boxes set on top of the roof. The metal canopies used to close the upper parts offer greater security when desired, largely depending on the seasonal conditions.

The architecture avoided the creation of a formal and functional connection with the existing structures. The living spaces that were planned and executed are flowing, free of framing elements, and thus generative of a greater freedom of use. An external staircase provides entrance to a vestibule. From there, access to each quarter is similarly incorporated. Each of the areas contains a living room plus a dining room integrated with the kitchen. The bedroom and bath follow. The upper story, with access to the rooftop terraces, holds a space that may be adapted to suit different needs.

The interiors contrast greatly with the cold look of the façade (largely the result of the metal/glass). The inner warmth comes from different choices of materials: the wooden floors and the predominant white tones inside the rooms creates a neutral feel open to warmer additions to taste. The inner staircases, of metal finished in white paint, serve as spatially defining elements with a sculptural touch. The construction of these stairs, which are in contact with other surfaces only at each end, again underlines the lightness that predominates throughout the whole project.

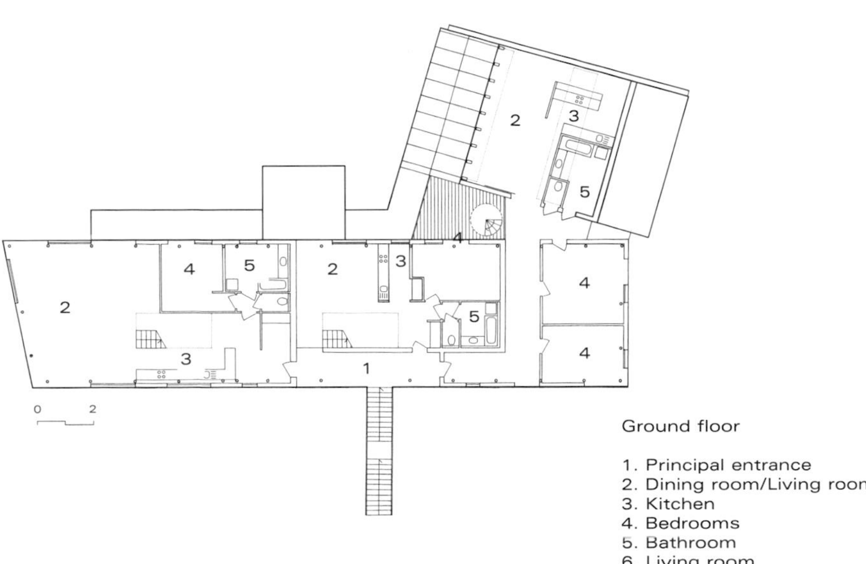

Ground floor

1. Principal entrance
2. Dining room/Living room
3. Kitchen
4. Bedrooms
5. Bathroom
6. Living room
7. Terrace

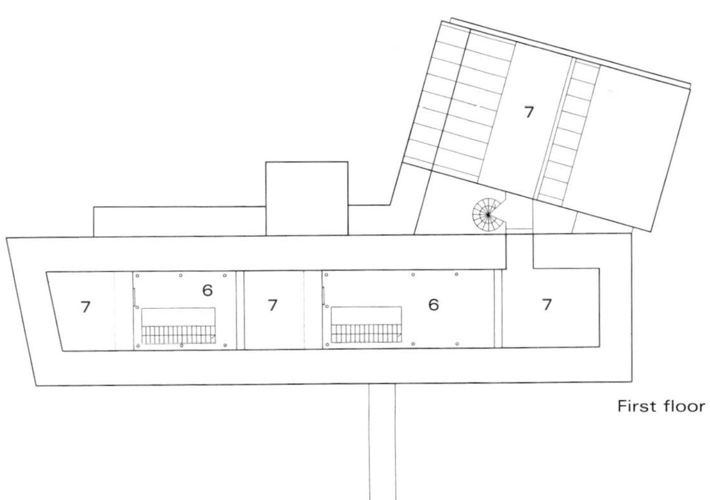

First floor

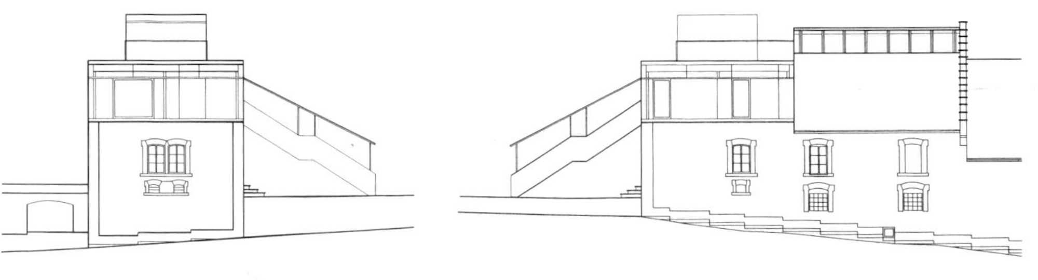

Sections

The regular rectangular shape and the building's perimeter metal structures made for the creation of well-defined spaces. These spaces, may, however, be redecorated for spatial variation. The relation between the different levels is brought out well by the predominance of the building's height, the staircase, and the great amount of natural light available.

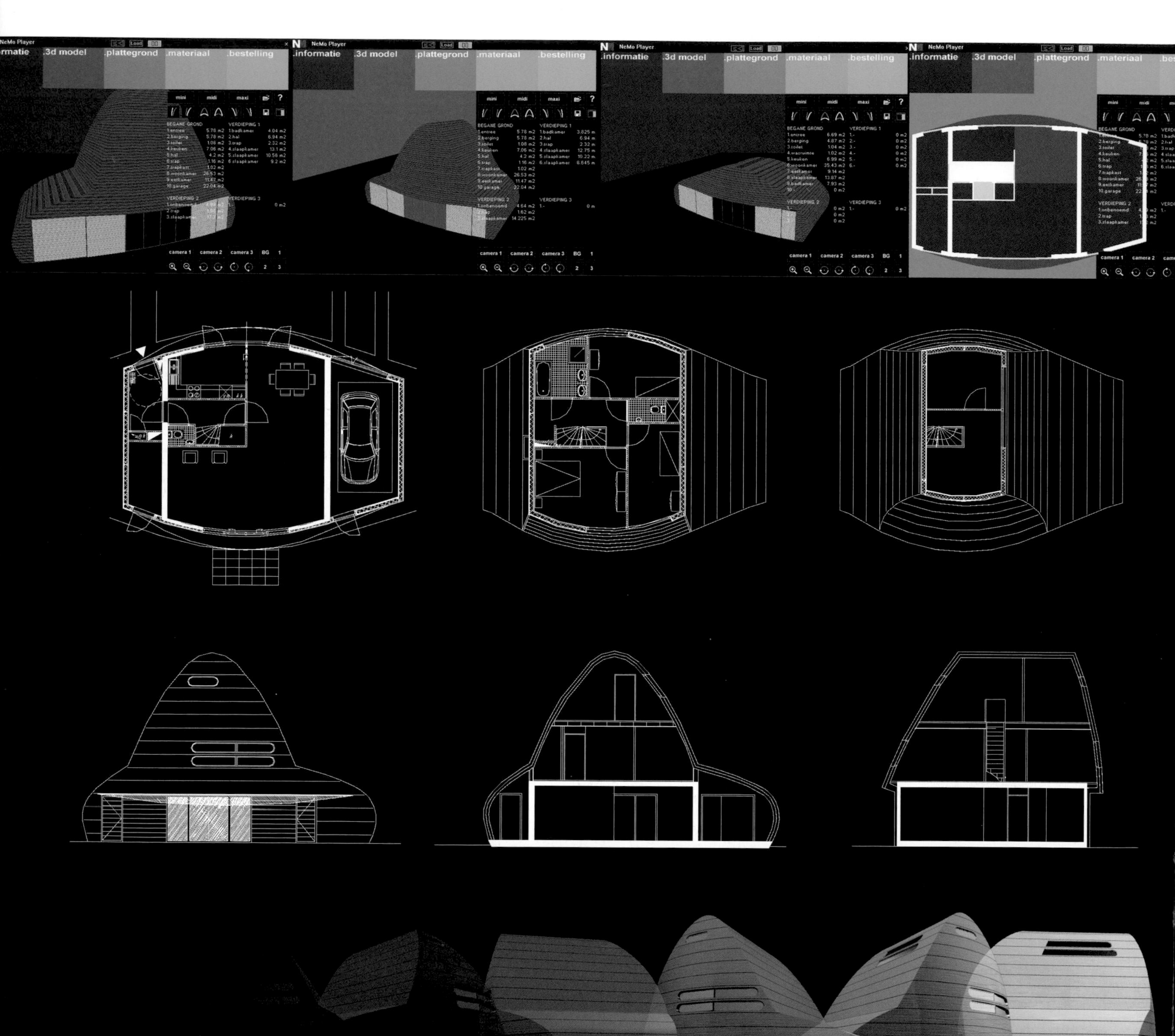
NeMo Player
Load
.informatie
.3d model
.plattegrond
.materiaal
.bestelling
mini
midi
maxi
?
BEGANE GROND
1.entree 5.78 m2
2.berging 5.78 m2
3.toilet 1.08 m2
4.keuken 7.06 m2
5.hal 4.2 m2
6.trap 1.16 m2
7.trapkast 1.02 m2
8.woonkamer 26.53 m2
9.eetkamer 11.47 m2
10.garage 22.04 m2
VERDIEPING 1
1.badkamer 4.04 m2
2.hal 6.94 m2
3.trap 2.32 m2
4.slaapkamer 13.1 m2
5.slaapkamer 10.56 m2
6.slaapkamer 9.2 m2
VERDIEPING 2
1.onbenoemd
2.trap
3.slaapkamer
VERDIEPING 3
0 m2
camera 1
camera 2
camera 3
BG
1
2
3

Architect: **Kas Oosterhuis / oosterhuis.nl** Location: **Anywhere**
Surface area: **323 sq. feet** Construction date: **2001**

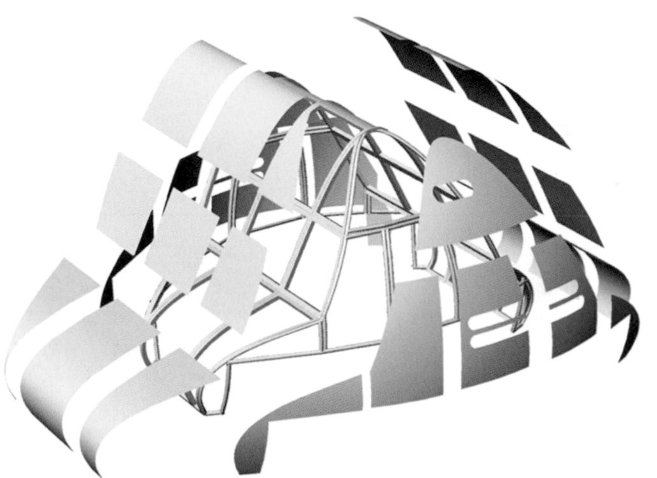

Variomatic

The concept of prefabricated construction brought with it the systematization of architecture and the possibility of cataloguing design schemes to allow their adjustment to new technical possibilities. This is is a concept of cataloguing housing projects that can be elastic on all quarters: height, depth, and width. Hence the name Variomatic, signifying the basic premise. The idea was for clients to participate actively in the design of their own house.

Such decision making would not only involve the final form of the curves in the façade or the roofs, but would also influence the dimensions of the building, the place where the kitchen should be, or the installation of a solar-energy water heater. By the same token, they could choose many of the materials and colors for the facings and the interiors.

By means of a systematic communication between client and architect and architect and builder, and thanks to the advanced methods of building, it is possible to construct a prototype very rapidly and still keep up the unity of the house. Since clients become participants in the project's design, Variomatic turns into a model where mass production can also achieve a varied scheme, changing according to specific needs.

www.variomatic.nl is a catalogue of prefab houses where clients become the designers of their own houses.

Mobility and the way the portable piece can be installed with minimal impact on the ecosystem creates an attractive counterpoint to the traditional home. This kind of architecture is redolent of times when the parts that made up a dwelling were easily manipulated and could be adapted to a very wide variety of conditions.

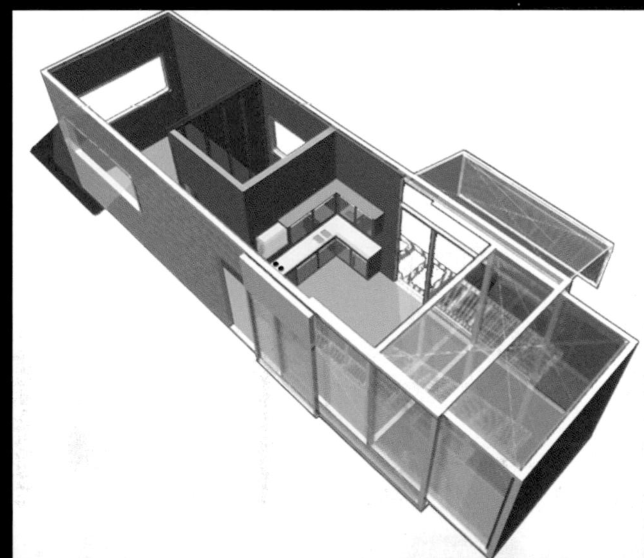

Architect: **Jennifer Siegal / Office of Mobile Design** Collaborators: **Elmer Arco, Thao Nguyen, Jon Racek.** Location: **Mobile** Surface area: **377 sq. feet (extendable)** Construction date: **2001**

Portable House

The Portable House may be adapted, extended, or moved to another site to meet new demands in a continually changing environment. The product thus offers a common-sense economical and ecological alternative to traditional houses and their its constantly growing expenses, which is to say, the majority of the options available today.

Portable House simultaneously returns to and criticizes preconceived notions of the mobile home and the trailer park. The project creates an entirely new option for those who have a limited budget to invest in a home but who don't have enough to make their way into the conventional home market.

It is a project that has been designed from a model of spaces that can be added to or subtracted from, as need dictates. By the same token, the degree of transparency of its materials or the relation of its internal spaces to the outdoors can be graduated, thus making it a really flexible creation. The kitchen and bath spaces, in the heart of the basic module, divide the sleeping space of the living and dining areas. The living room structure can be extended in two directions if it is necessary to use more interior.

The concept allows the house to be installed in different positions, thus taking advantage of the best orientation in changing climatic conditions. On the other hand, some models can be assembled together to create a varied typology of architecture and relationships with the natural environments.

■ Duimdrop
Joost Glissenaar & Klaas van der Moelen / BAR
Pelgrimsstraat 5-B, 3029BH Rotterdam, Holland
Tel.: +31 (0)10 477 3863
Fax: +31 (0)10 476 6615
BAR.glism@zonnet.nl

■ Kielder Belvedere - Floating Island - Tree House
SOFTROOM ARCHITECTS
34 Lexington Street, Londres W1F OLH, United Kigdom
Tel.: +44 (0)20 7437 1550
Fax: +44 (0)20 7437 1566
softroom@softroom.com
www.softroom.com

■ Modular Dwellings
Edgar Blazona
2729 Acton Street, Berkley, CA 94702 U.S.A.
Tel.: +1 415 581 3321
EBlazona@WSGC.com

■ Black Maria
Hiroshi Nakao
Edogawa Apt. 511, 6-18 Shin-Ogawa, Shinjuku, Tokyo, 162 0814, Japan
Tel.: +81 3 3235 2902
nasa@gw4.gateway.ne.jp

■ Cliff House
Meindert Versteeg
Daam Fockemlaan 20, 3818 KG Amersfoort, Holland
Tel.: +31 33 461 7274
Fax: +31 65 217 5401
meinderstversteeg@wanadoo.nl

■ iMobile + Portable House
Jennifer Siegal / OFFICE OF MOBILE DESIGN
642 Moulton Avenue Studio E34 Los Ángeles, CA 90031, U.S.A
Tel.: +1 323 441 9776
Fax: +1 323 441 9775
jennifer@designmobile.com
www.designmobile.com

■ Yardbird Studio
NEAL R. DEPUTY ARCHITECT INC.
520 Lincoln Road, Miami Beach, FL 33139-7733, U.S.A
Tel.: +1 305 534 4020
Fax: +1 305 534 4095
nrdai@aol.com

■ Keroman Nautical Base
Jean-François Revert
18 Passage du Chantier 75012 París, France
Tel.: +33 1 43 44 11 25
jf-revert@worldnet.fr

■ Oasis Apartments
Hans Peter Wörndl, Wolfgang Tschapeller, Max Rider
Dornbacherstrasse 107, 1170 Viena, Austria
Tel.: +43 1 48 61 102
Fax: +43 1 48 61 102
woerndl@nusurf.at

■ Doppelhofer +
Spieldfeld Border Control + Stockner House
Wolfgang Feyferlik
Glacisstrasse 7, 8010 Graz, Austria
Tel.: +43 316 34 76 56
Fax: +43 316 38 60 29
feyferlik@inode.at

■ House in Higashi-Osaka
Waro Kishi
3F Yamashita Bldg. 10 Nishimotomachi, Koyama, Kita-ku, Kyoto 603 Japan
Tel.: +81 75 492 5175
+81 75 492 5185
kishi@k-associates.com

■ Bus Stop
Michael Culpepper + Greg Tew
1220 Boren Avenue # 708, Seattle, WA 98101
U.S.A
Tel.: +1 206 521 3509
Fax: +1 206 623 7868
mcworkshop@aol.com

■ Manufactured Housing
Andrew Thurlow + Maia Small / TSA ARCHITECTS
3708 Timberlake Drive, Knoxville, TN 37920
U.S.A
Tel.: +1 865 386 5684
maia@utk.edu

■ Mailand House
Oskar Leo Kaufmann
Steinebach 3, 6850 Dornbirn, Vorarlberg, Austria
Tel.: +43 5572 39 49 69 – 0
Fax: +43 5572 39 49 69 –20
office@olk.cc
www.olk.cc

■ Pavilion of Yamaguchi Prefecture
Katsufumi Kubota
1-8-24 Imazu-cho, Iwakuni, Yamaguchi 740-0017 Japan
Tel.: +81 827 22 0092
Fax: +81 827 22 0079
kubotaaa@ymg.urban.ne.jp
www.urban.ne.jp/home/kubotaaa

■ Expo'98 Kiosks
Joâo Mendes Ribeiro + Pedro Brígida
Rua Alexandre Herculano, 16B, 1°, 3000-019 Coimbra, Portugal
Tel.: +35 1 239 833 763
Fax: +35 1 239 833 763
mribeiro@interacesso.pt

■ Servicenter
STEINMANN & SCHMID ARCHITEKTEN
Rebgasse 21A CH 4058 Basel, Switzerland
Tel.: +41 61 686 9300
Fax: +41 61 686 9301
www.steinmann-schmid.ch

■ GucklHupf
Hans Peter Wörndl
Dornbacherstrasse 107, 1170 Vienna, Austria
Tel.: +43 1 48 61 102
Fax: +43 1 48 61 102
woerndl@nusurf.at

■ Kosovo Kit
Jeremy Edmiston, Douglas Gauthier / SYSTEM ARCHITECTS
249 West 29th Street Nº2N, Nueva York, NY 10001, U.S.A
Tel.: +1 212 239 8001
Fax: +1 212 239 8005
system@systemarchitects.net
www.systemarchitects.net

■ Burghalde
ALIOTH LANGLOTZ STALDER BUOL
Neugasse 6, 8005 Zurich, Switzerland
Tel.: +41 1 273 33 39
Fax: +41 1 273 33 38
leobuol@alsb-arch.ch

■ Kindergarten in Pliezhausen
D'INKA, SCHEIBLE + PARTNER FREIE ARCHITEKTEN
Bahnhofstrasse 52 70734 Fellbach, Germany
Tel.: +49 711 57 5888
Fax: +49 711 580 784
architekten-dinka-scheible@t-online.de

■ Naestved Pedestrian Bridge
ANDERSEN & SIGURDSSON ARKITEKTER
Vestergade 10B, 2. sal, 1456 Kobenhavn, Denmark
Tel.: +45 3369 00 85
Fax: +45 3369 00 36
andersen.sigurdsson@get2net.dk

■ Tahquitz Canyon Visitors Center
O'DONELL + ESCALANTE ARCHITECTS
121 S. Palm Canyon Dr. Ste. 222, Palm Springs, CA 92262
U.S.A
Tel.: +1 760 323 1925
Fax: +1 760 320 7897
oearch@aol.com

■ G House + Hornegg Complex
Hans Gangoly
Volksgartenstrasse 18, 8020 Graz, Austria
Tel.: +43 316 71 75 50
Fax: +43 316 71 75 56
office@gangoly.at

■ Pen House
QUERKRAFT ARCHITECTS
Pilgramgasse 1/2/17 a-1050 Vienna, Austria
Tel.: +43 (0)1 548 7711
Fax: +43 (0)1 548 7744
office@querkraft.at
www.querkraft.at

■ Tea House
MARTERERMOOSMANN
Grinzingerallee 50-52/6, 1190 Vienna, Austria
Tel.: +43 1 32 89 270
Fax: +43 1 32 89 220
office@marterermoosmann.com
www.marterermoosmann.com

■ Wooden Modules
Arthur Collin
la Berry Place, Londres ECIV 0JD, United Kingdom
Tel.: +44 (0)20 7490 3520
Fax: +44 (0)20 7490 3521

■ Modular Shelters
JONES, PARTNERS: ARCHITECTURE
141 Nevada Street, El Segundo, CA 90245, U.S.A
Tel.: +1 310 414 0761
Fax: +1 310 414 0765
wes@jonespartners.com

■ Subway Station
DESPANG ARCHITEKTEN
Am Graswege 5, 30169 Hannover, Germany
Tel.: +49 511 88 28 40
Fax: +49 511 88 79 85
Despang@BauNetz.de
www.DespangArchitekten.de

■ Two Family House
Oskar Leo Kaufmann and Bmst Johannes Kaufmann
Steinebach 3, 6850 Dornbirn, Vorarlberg, Austria
Tel.: +43 5572 39 49 69 – 0
Fax: +43 5572 39 49 69 –20
office@olk.cc
www.olk.cc

■ House in Munich
Christof Wallner
Nymphenburger Straβe 47, 80335 Munich, Germany
Tel.: +49 89 127 00 55 – 5
Fax: +49 89 127 00 55 – 7
ch.wallner@t-online.de

■ Variomatic
OOSTERHUIS.NL
Essenburgsingel 94c, nl-3022 eg Rotterdam, Holland
Tel.: +31 10 244 7039
Fax: +31 10 244 7041
oosterhuis@oosterhuis.nl
www.osterhuis.nl